MW01385787

# Risk, Systems and Decisions



More information about this series at http://www.springer.com/series/13439

Matthew D. Wood • Sarah Thorne
Daniel Kovacs • Gordon Butte • Igor Linkov

# Mental Modeling Approach

## Risk Management Application Case Studies

Matthew D. Wood
U.S. Army Corps of Engineers
Carnegie Mellon University
Concord, MA, USA

Sarah Thorne
Decision. Partners Inc.
Decision Partners, LLC
Mississagua, ON, CAN

Daniel Kovacs
Decision Partners, LLC
Pittsburgh, PA, USA

Gordon Butte
Decision Partners, LLC
Pittsburgh, PA, USA

Igor Linkov
U.S. Army Corps of Engineers
Carnegie Mellon University
Concord, MA, USA

Risk, Systems and Decisions
ISBN 978-1-4939-6614-1 ISBN 978-1-4939-6616-5 (eBook)
DOI 10.1007/978-1-4939-6616-5

Library of Congress Control Number: 2016951930



Printed on acid-free paper

This Springer imprint is published by Springer Nature
The registered company is Springer Science+Business Media LLC
The registered company address is: 233 Spring Street, New York, NY 10013, U.S.A.

*Mental Modeling is dedicated to …*
*Baruch Fischhoff, for teaching us, inspiring us, and supporting the application of mental modeling research to solve real-world problems;*
*The Department of Engineering and Public Policy, Carnegie Mellon University, which long ago recognized the need for multidimensional approaches to complex challenges and focused its efforts on teaching and graduating smart, passionate systems thinkers. We are grateful to the professors and graduate students with whom we have, and continue to have, the opportunity to collaborate;*
*Our many colleagues and partners, who have and continue to support us in pursuit of science-informed, evidence-based solutions based on Mental Modeling; and*
*Last, but not least, our staff and research colleagues at Decision Partners and U.S. Army Corps of Engineers, whose hard work, sharp minds, and dedication are woven into the fabric of every chapter.*
*We hope Mental Modeling is a tribute to all and an inspiration for systems thinkers and problem solvers around the world.*

*The authors*

# Foreword

Effective risk communication requires contributions from subject matter experts, who know the issues; analysts, who can identify the essential ones; behavioral scientists, who can address audience members' information needs; and specialists, who can create channels for trusted two-way communication between the parties. The mental models approach provides a framework for organizing the information needed to accomplish this task. However, it takes deep personal and organizational commitment to bring and keep the parties together. *Mental Modeling* shows how to make that happen, integrating theory and practice.

The range of its applications is remarkably broad, including plastic surgery, climate change, dairy farming, deep mining, biosolids, nuclear power, and carbon capture and sequestration. So is the range of stakeholders and audiences, including physicians, patients, regulators, laborers, engineers, land use planners, and river managers. And, so are the methods, including community workshops, in-depth interviews, expert elicitation sessions, computer models, worker training, and broad and narrowband communication. These ranges of topics, audiences, and method show the generality of the approach and the creativity of the authors in its use.

Readers of *Mental Modeling* will acquire an understanding of the theory underlying the approach, with its basic principles illustrated in diverse, practical examples. Readers will learn methods that they can apply directly and strategies for generating their own. And they will come away with an appreciation of the diligence needed to create communications worthy of the stakes riding on them. Although not easy, the work is exciting—and gratifying.

Baruch Fischhoff, PhD
Pittsburgh, PA

Dr. Fischhoff is Howard Heinz University Professor with the Departments of Social and Decision Science and of Engineering and Public Policy at Carnegie Mellon University, and Decision Partners' Chief Scientist. Dr. Fischhoff is a Fellow of the American Psychological Association and a Fellow of the Society of Risk Analysis, as well as recipient of its Distinguished Achievement Award (1991). He is a member of the Institute of Medicine (IOM) and of the National Academies and has participated in some two dozen committees of the IOM and National Research Council. Baruch is a coauthor of several books including *Acceptable Risk (1981), Risk Communication—a Mental Models Approach (2002),* and *Intelligence Analysis: Behavioral and Social Science Foundations (2011).* He has also coauthored *Communicating Risks and Benefits: An Evidence-Based User's Guide (2011).* He holds a PhD in psychology from the Hebrew University of Jerusalem.

## Perspective on *Mental Modeling*

Throughout the U.S. Army Corps of Engineers (USACE), leaders at all levels and across all mission areas face increasingly complex demands. Projects are more technically challenging than ever before, regulatory requirements are more difficult, economic pressures are greater, and the universe of stakeholders is broader and more engaged than even a decade ago. At USACE's Engineer Research and Development Center's (ERDC's) Environmental Lab, one of our key charges is to develop, test, and disseminate practical tools and methods throughout USACE, including those designed to better align and integrate ecological, engineering, and social sciences considerations that result in more socially acceptable, economically viable, and environmentally sustainable projects. The stakes are high for the USACE's activities, as all projects USACE undertakes are done through different degrees of collaboration with agency partners and key stakeholders, with potential for both positive and negative impacts on local ecology and environment, socioeconomic health of the community and region, etc. As we have learned, the quality of our stakeholder engagement processes from project design through implementation affects both the efficiency and quality of project decision making and, often, project success or failure.

The USACE's typical agency partners and external stakeholders are wide ranging and include a number of other federal agencies. Often lead agency partners have overlapping, or sometimes conflicting, regulatory authorities. They and other stakeholders may have competing objectives, interests, values, and priorities. "Social friction" arises in the planning process when different agency and key stakeholder perceptions, goals, values, and capacities lead to different judgments about a proposed project's value (Chap. 10). This means that partner and stakeholder interaction is often difficult, complicated, highly scrutinized, and under pressure due to lack of alignment on goals and desired outcomes. A recent internal USACE assessment documented the need for better, more flexible stakeholder collaboration processes, more internal training, and ready access to resources and specialized skills. Identified concerns included perceptions that some stakeholders believe they are engaged too late for their input to be valued in decision-making processes, that their input is not valued, and that USACE is not really concerned about the environment. These difficulties increase completion time and operational costs of infrastructure projects. Lack of clear process, increasing time constraints, and diminishing financial and human resources within USACE and agency partners compound these challenges.

The need to find and apply science-informed, evidence-based stakeholder engagement and communication processes in order to take into account the varying goals, values, and priorities of the many stakeholders with an interest in a USACE project led us to explore Mental Modeling Technology™ over a decade ago. The Mental Modeling approach starts with engaging experts to develop a system model, or *expert model*. An expert model is a formal, comprehensive graphic representation that summarizes and integrates the current knowledge and understanding of experts about the key factors of the topic being studied. It can be thought of as an *expert's mental model*, as it typically comprises a composite of the knowledge and beliefs—mental models—of several experts. That model then serves as the foundation to systematically engage a wide range of stakeholders through formal or informal research. This approach has provided the base for developing a number of initiatives at ERDC, in collaboration with other USACE colleagues. Examples of challenges that have benefited from this approach are described in the chapters that follow.

We have found for complex topics, especially those where the science is uncertain or incomplete, bringing together experts in a workshop setting and using Mental Modeling tools and techniques to elicit a broad range of expertise and experience is highly beneficial. In the case studies that follow on Flood Risk Management (Chap. 4) and Adaptive Management for Climate Change, as well as our work on beneficial use of dredged material (Chap. 5), we did just that. In each case, the focused expert elicitation resulted in the development of a comprehensive system picture, or expert model, which was then validated with the respective expert participants. Not only did this approach build shared understanding of the system and the critical influences on the desired outcomes, but it also served as a focal point for bringing diverse experts from across USACE, along with those from agency partner and stakeholder organizations, together to share insight and expertise on the subject matter at hand in a neutral forum. The resulting models were then used to establish strategic priorities, research agendas, and, in the case of Flood Risk Management (Chap. 4), the analytical framework for the follow on *mental models* research.

The application of Mental Modeling to Technology Infusion and Marketing (Chap. 6) was a different application and one of major significance to ERDC. A critical challenge for any research organization, including ERDC, is the ability to get new technology out of the lab and applied in the field. With pressure on budgets, time, and resources, this challenge was increasingly becoming a barrier. The Mental Modeling approach was used to first understand the current situation for technology transfer and adoption, then to develop, with USACE stakeholders, a recommended Technology Infusion and Marketing approach, along with the critical success criteria. This streamlined approach is producing results for the Environmental Lab and beyond.

We continue to apply the Mental Modeling process, methods, and tools to a range of complex challenges across the USACE's mission areas. With our first applications in Navigation and Flood Risk Management, we've since broadened its application across Civil Works and, to a small degree, our Military mission, bringing our internal and external stakeholders together to solve multidimensional problems using this integrated approach.

Perhaps the strongest case for the Mental Modeling approach is it application in the design, implementation, and measurement of our groundbreaking Engineering with Nature (EWN) initiative. Since early 2011, a core team of scientists and engineers at ERDC have been applying the fundamental concepts and approaches of Mental Modeling to develop EWN. Collaboration with key internal USACE stakeholders and with external agency partners and stakeholders was a critical component of its design. Now a USACE Program, EWN represents a paradigm shift from USACE's traditional decision-making model, perceived by some agency partners and stakeholders as confrontational, to one of more effective decision making through early and ongoing collaboration with partners and stakeholders. And as the demonstration projects are showing, it produces triple win results, typically faster, more efficiently, and without the social friction typical of many previous USACE-led projects. EWN is seen as enabling transformation across USACE and beyond, with and through the Corps' agency partners and stakeholders. It is noteworthy that EWN was recently recognized with two awards, USACE 2014 Green Innovation Award and Western Dredging Association (WEDA) 2015 Environment Award, and the publication of the North Atlantic Coast Comprehensive Study (NACCS) Natural and Nature-Based Features Report.

Our application of Mental Modeling continues to add value across USACE and beyond as we apply it to a range of increasingly complex challenges, while building our skills and stakeholder engagement capacity in the process.

Todd Bridges, PhD
Vicksburg, MS

Todd Bridges is Senior Research Scientist, U.S. Army Corps of Engineers (USACE), Engineer Research & Development Center (ERDC). He currently leads USACE's Engineering with Nature Initiative, which includes a network of research projects, field demonstrations, and communication activities to promote environmentally sustainable infrastructure development. He has chaired international working groups for the London Convention and Protocol which have developed technical guidance for assessing sediments as well as managing risks associated with $CO_2$ sequestration operations in the oceans. As U.S. representative to the Environmental Commission of the International Navigation Association (PIANC), Dr. Bridges has led efforts to develop new international standards for managing environmental risks, while promoting environmental benefits, related to navigation infrastructure.

He has served on the editorial boards of the journals of *Integrated Environmental Assessment and Management, Environmental Toxicology and Chemistry,* and *Dredging Engineering*. He is an active member of the Society for Risk Analysis, The Society of Environmental Toxicology and Chemistry, The Ecological Society of America, and the International Navigation Association.

Over the last 20 years, Dr. Bridges has published more than 60 journal articles and book chapters and numerous technical reports. He received his B.A. (1985) in Biology/Zoology from California State University, Fresno, and his PhD (1992) in Biological Oceanography at North Carolina State University.

## From a Practitioner's Perspective

Prior to first hearing about Mental Modeling, I had spent several years developing and implementing reputation management processes for various private industry organizations. I was working on a new, large-scale project that I knew would be challenging—challenging not just because we were building significant energy infrastructure in farm country but also because we were dealing with many different stakeholder viewpoints. Recognizing that we as an organization needed to change the way we approached infrastructure development, I was searching for a tool that would take into account the values and interests of engineers, business people, landowners, environmentalists, and government, find the alignments among all these stakeholders and, based on that, enable us to develop a respectful, collaborative process. A chance meeting with Decision Partners at an industry event led me to that tool—Mental Modeling Technology™.

Since then, I have used Mental Modeling not only in stakeholder consultation on infrastructure development but also in strategic plan development for industry associations undergoing intense change. In each instance, the systematic, science-based Mental Modeling approach enabled us to dive deeply into the thinking of a range of stakeholders and truly understand what's in their hearts and minds, and what forms that.

The Mental Modeling interviewing process is a very deep process that gets not only at what people think and believe but why they think and believe it. Having this insight enables the industry practitioner to identify trends in these thoughts and beliefs as to how they influence stakeholder judgment. To me and many of my colleagues, this critical insight is what makes Mental Modeling the ideal strategic tool to formulate an appropriate consultation strategy and respectful dialogue with stakeholders that enables them to participate in the decision-making process in a way that is meaningful to them.

Ultimately, it doesn't matter whether you're a corporate CEO, a government person, or a landowner. It's about respecting stakeholder beliefs and values and working within the confines of those beliefs and values so that you understand all of those different stakeholders.

My advice to other private industry practitioners would be: if you're looking for a truly systematic and science-based approach to understand the decision making of both industry leaders and their key stakeholders, Mental Modeling offers great value.

Denise Carpenter
Toronto, ON

Denise Carpenter is Chief Executive Officer of the Neighbourhood Pharmacy Association of Canada. Her diverse leadership experience spans key economic sectors and organizations. As an applied management and social sciences professional, Denise has used Mental Modeling in varied organizational contexts: corporate governance, strategic planning and implementation, systematic behavior change, integrated risk management, public policy and public affairs. She also has expertise in leadership development, change management, and culture change through innovative communications.

# Preface

## Purpose of the Book

The goal of this book is to introduce readers to *Mental Modeling*, an evidence-based process to facilitate decision making by describing the values and knowledge of individuals involved in the decision-making process. The book is tailored to students and practitioners in environmental and risk management domains who have some experience with the complex, often difficult projects that require engagement and understanding of the thoughts and beliefs of different stakeholder groups. Mental Modeling is ideally suited to contexts in which (a) the issues of interest are complex with a significant degree of consequence, (b) disparate viewpoints related to the issue or opportunity gaps must be synthesized, (c) decisions are required among multiple potential alternative risk management options, and/or (d) transparency is required when characterizing the issue, incorporating stakeholder input, designing appropriate risk management solutions, and justifying risk management actions. It is particularly well suited to identifying relationships among influences that may not be easily anticipated and providing a basis for developing or comparing solution alternatives for complex real-world problems.

## How to Read This Book

This book is designed as an introduction to students and practitioners in public policy, risk communication, and related disciplines. The first sections provide an introduction to the process historically and as it stands today, and should be reviewed first to provide context to the other content in the book. Subsequent chapters, in contrast, are intended as a showcase of the different application domains where Mental Modeling has been successfully applied to address complex problems. These chapters can be read piecemeal depending on the reader's needs and interests. Commentaries and testimonials are dispersed throughout the text to highlight some of the method's strengths and future directions for using the Mental Modeling approach.

Concord, MA, USA — Matthew D. Wood
Mississagua, ON, CAN — Sarah Thorne
Pittsburgh, PA, USA — Daniel Kovacs
Pittsburgh, PA, USA — Gordon Butte
Concord, MA, USA — Igor Linkov

# Acknowledgements

The authors would like to sincerely thank and acknowledge the following colleagues, client partners, and friends for their contributions to this book, as well as many others who have contributed to the development of the Mental Modeling approach:

Steve Ackerlund, PhD, Senior Scientist and Research Lead, Decision Partners
Kelsie Baker, former intern with U.S. Army Corps of Engineers
Anne Bostrom, PhD, Weyerhaeuser Endowed Professor in Environmental Policy, University of Washington, Daniel J. Evans School of Public Affairs
Todd Bridges, PhD, Senior Research Scientist, Environmental Sciences, U.S. Army Engineer Research & Development Center
Denise Carpenter, President and Chief Executive Officer, Neighbourhood Pharmacy Association of Canada
Florence Chang, Branch Chief, National Library of Medicine
Tanya Darisi, PhD, Director, The O'Halloran Group and former Senior Researcher with Decision Partners
Gia DeJulio, Director, Regulatory Affairs, Enersource Hydro Mississauga
Sara Eggers, PhD, Economist, U.S. Food and Drug Administration and former Research Director with Decision Partners
Baruch Fischhoff, PhD, Howard Heinz University Professor, Social and Decision Sciences and Engineering and Public Policy, Carnegie Mellon University, and Decision Partners' Chief Scientist
Elizabeth Fleming, PhD, Director of Environmental Laboratory, U.S. Army Engineer Research & Development Center
Robert Green, Principal, Penn Schoen Berland
Samuel Hagedorn, Director, Penn Schoen Berland
Sarah Hailey, PhD, Research Designer and Analyst, Institutional Research and Analysis, Carnegie Mellon University, and former Research Scientist with Decision Partners
Alan Hais, Retired Program Director, U.S. Water and Environment Foundation (WERF)

Pertti Hakkinen, PhD, Acting Head, Specialized Information Services Office of Clinical Toxicology, National Library of Medicine
Martine Hartogensis, DVM, Deputy Director, Office of Surveillance & Compliance, Health and Human Services
Martin Kennedy, Vice-President, External Affairs, Capital Power Corporation
Carlton Ketchum, Creative Operations Director, MedRespond
Nick Martyn, Founder and CEO, RiskLogik
Kevin Miller, Principal/Partner, Reingold
Joseph Ney, Creative Director, Principal, Reingold
Jennifer Pakiam, Technical Information Specialist, National Library of Medicine
Virginia Pribanic, President, MedRespond, LLC
Katherine Sousa, Senior Researcher, Decision Partners
John Stewart, PhD, Principal, J M Stewart Consulting, South Africa
Alexander Tkachuk, Software Architect/Developer, Decision Partners
Monica Wroblewski, PhD, Researcher, Communications Directorate, U.S. Census Bureau
Megan Young, Senior Consultant with Decision Partners
We would also like to thank the following organizations for their support with several of our Mental Modeling applications:
American Society of Plastic Surgeons
Capital Power Corporation
Chamber of Mines of South Africa
U.S. Census Bureau
Enersource Hydro Mississauga
EPCOR Utilities Inc.
U.S. Food and Drug Administration
U.S. Health and Human Services
National Library of Medicine
Penn Schoen Berland
Reingold
Technical Standards and Safety Authority of Ontario
U.S. Army Corps of Engineers
U.S. Water and Environment Research Foundation

Special thanks to Anne Papmehl, Senior Writer, Editor and Researcher, Decision Partners, for her editorial support which was critical to the completion of *Mental Modeling* and to Melinda Paul at Springer, whose interest and support made this book possible.

# Contents

# List of Abbreviations

| | |
|---|---|
| ACGIH | American Conference of Government Industrial Hygienists |
| ACS | American Community Survey |
| AIHA | American Industrial Hygiene Association |
| APHL | Association of Public Health Laboratories (APHL) |
| ASA(ALT) | Office of the Assistant Secretary of the Army for Acquisition Logistics and Technology |
| ASPS | American Society of Plastic Surgeons |
| ATSDR | Agency for Toxic Substances and Disease Registry |
| BEA | Bureau of Economic Analysis |
| BLS | Bureau of Labor Statistics |
| CAMEO | Computer-Aided Management of Emergency Operations |
| CARP | Canadian Association of Retired Persons |
| CASS™ | Cognitive Analysis Software Suite |
| CB | U.S. Census Bureau |
| CBRNE | Chemical biological, radiological, nuclear and explosive emergencies |
| CCS | Carbon Capture and Storage |
| CDC | Center for Disease Control |
| CDC Epi-X | CDC Epidemic Information Exchange |
| CHEMM | Chemical Hazards Emergency Medical Management Tool |
| CHEMTREC | CHEMical Transportation Emergency Center |
| CMSA | Carbon Monoxide Safety Association |
| CO | Carbon monoxide |
| CoP | Community of Practice (also COPA) |
| COPA | Community of Practice (also CoP) |
| CRAR | Communications Research and Analytics Roadmap |
| CSA | Canadian Standards Association |
| CVM | Center for Veterinary Medicine (FDA) |
| DHHS | Department of Health and Human Services |
| DHS | Department of Homeland Security |
| DOD | Department of Defense |
| DOJ | Department of Justice |

| | |
|---|---|
| DOT | Another name for Emergency Response Guidebook (ERG) and sometimes called Orange Response Guidebook |
| EIA | Environmental Impact Assessment |
| EL | Environmental Laboratory |
| EMT | Emergency Medical Technician |
| EPA | U.S. Environmental Protection Agency |
| ER | Emergency Room |
| ERDC | Engineering Research and Development Center |
| ERG | Emergency Response Guidebook |
| ERM | Environmental Risk Management |
| EWN | Engineering with Nature |
| FBI | Federal Bureau of Investigations |
| FDA | (U.S.) Food and Drug Administration |
| FEMA | Federal Emergency Management Agency |
| FWS | U.S. Fish and Wildlife Service |
| GAO | Government Accountability Office |
| HazMat | Hazardous Materials |
| HHS | Department of Health and Human Services |
| HHS SOC | HHS Secretary's Operations Center |
| HUD | Department of Housing and Urban Development |
| ICTW | Interstate Chemical Threats Workgroup |
| IGCC | Integrated Gasification and Combine Cycle |
| ILO | International Labour Organization |
| IPCC | Intergovernmental Panel on Climate Change |
| ISO | International Organization for Standardization |
| MOSH | Mining Industry Occupational Safety and Health |
| MSDS | Material Safety Data Sheets |
| NED | National Economic Development |
| NFF | National Fire Fighter Corp. |
| NFIP | National Flood Insurance Program |
| NFPA | National Fire Protection Association |
| NIH | National Institutes of Health |
| NIOSH | National Institute for Occupational Safety and Health |
| NLM | National Library of Medicine |
| NLM HSDB | NLM Hazardous Substances Data Bank |
| OH&S | Occupational Health & Safety |
| OIG | Office of the Inspector General |
| OSHA | Occupational Safety and Health Administration |
| PCB | Polychlorinated biphenyl |
| PHO | Public Health Official |
| PSB | Penn Schoen Berland |
| REMM | Radiation Emergency Medical Management |
| SIS | Specialized Information Services Division of National Library of Medicine |
| TIM | Technology Infusion and Marketing |

| | |
|---|---|
| TOR | Terms of Reference |
| TREECS | Training Range Environmental Evaluation and Characterization System |
| TSSA | Technical Standards and Safety Authority of Ontario |
| USACE | U.S. Army Corps of Engineers |
| VDH | Virginia Department of Health |
| WERF | Water and Environment Research Foundation |
| WISER | Wireless Information System for Emergency Responders |
| WMD-CSTs | Weapons of Mass Destruction Civil Support Teams |

# Chapter 1
# An Introduction to Mental Modeling

**Matthew D. Wood, Sarah Thorne, Daniel Kovacs, Gordon Butte, and Igor Linkov**

The goal of this book is to introduce readers to *Mental Modeling*, an evidence-based process to facilitate decision making by describing the values and knowledge of individuals involved in the decision-making process. Over the ensuing decades of application, Mental Modeling has been in continuous development, extending from initial applications in risk communication and risk management into a broad range of other applications, from strategic planning, to stakeholder engagement to change management and technology transfer and adoption.

M.D. Wood, Ph.D. (✉)
U.S. Army Corps of Engineers, Engineer Research and Development Center (ERDC) and Carnegie Mellon University, 696 Virginia Road, Concord, MA 01742, USA
e-mail: Matthew.d.wood@usace.army.mil

S. Thorne, M.A.
Decision Partners, 1084 Queen Street West, #32B, Mississauga, ON, Canada L5H 4K4
e-mail: sthorne@decisionpartners.com

D. Kovacs, Ph.D.
Decision Partners, 1458 Jersey Street, Lake Milton, OH 44429, USA
e-mail: dkovacs@decisionparters.com

G. Butte
Decision Partners LLC, Suite 200, 313 East Carson Street, Pittsburgh, PA 15217, USA
e-mail: gbutte@decisionpartners.com

I. Linkov, Ph.D.
U.S. Army Corps of Engineers, Engineer Research and Development Center (ERDC) and Carnegie Mellon University, 696 Virginia Road, Concord, Boston, MA 01742, USA
e-mail: Igor.Linkov@usace.army.mil



M.D. Wood et al., *Mental Modeling Approach*, Risk, Systems and Decisions,
DOI 10.1007/978-1-4939-6616-5_1

## Supporting Evidence-Based Strategies and Communications

Everyone is a decision-maker. Their decision problems vary widely from the personal (buying a new car) to the professional (deciding to move operations to a different part of the country). The decisions may be simple and trivial, or they may be complex, involving risks or opportunities, with potentially significant outcomes, making the stakes high for all who are involved or affected by these decisions.

Regardless of the decision, the fundamental cognitive task is the same. People must think through the options, the risks and benefits—pros and cons—consider the uncertainties and weigh the trade-offs as they go. Sometimes, the decision making can be fast and loose, requiring little conscious effort. Other times, it may require a slow and rigorous process, requiring careful, time-consuming deliberation, with a heavy, taxing cognitive workload.

To make decisions, people need a diverse set of cognitive, social, and emotional skills in order to understand and interpret the knowledge they have and the information they receive from communications of various kinds. Those skills can be acquired through formal education, self-study, and personal experience. No matter how skilled people are at making decisions, they often need trustworthy, accessible, and comprehensible information about the specifics of the risks, benefits, and trade-offs involved in order to make well-informed decisions and take appropriate actions, especially for the more complex and consequential decisions. Risk communication is a science-informed, evidence-based approach to providing the relevant and useful information people need to make sound choices in such situations.

Risk communication, as a practice, is distinguished from other communication approaches by the commitment of its practitioners to its purpose—enabling people to make well-informed decisions in complex, consequential situations. It is predicated on the need to provide accurate, useful information in a balanced and unbiased way. As an integral component of effective risk management, risk communication must deal with communicating both the benefits that decisions can produce, such as profits from investments, or improved health from medical procedures, as well as the risks, such as bankruptcy from a failed start up or side effects from a medical procedure. People need to understand both in order to fairly weigh the options and make an appropriate decision.

The academic research literature on risk communication is large and diverse, representing a wide range of applications and drawing on many contributing disciplines, such as psychology, decision science, sociology, and communications. Unfortunately, academic research in these fields is often unaccessed by risk communication practitioners, creating significant barriers to adoption of state-of-the-science knowledge in practical application. As a result, some risk communications practitioners rely on intuition, personal experiences, and popular interpretations of psychological research, to employ so-called best practices, rather than proven, science-based approaches.

Over the past two decades, a network of management professionals, researchers, and social scientists have developed a science-informed management process for risk communication. It fills the gap, effectively bridging risk communication science to the practical challenges practitioners face as they work to better inform judgment, decision making, and behavior. The methodology has come to be called Mental Modeling. Its roots, and the term, trace back to the 1930s with the first description of the cognitive phenomenon of mental models. Foundational work in mental models-based risk perception and risk communication led by Dr. Baruch Fischhoff[1] and his colleagues at Carnegie Mellon University in the 1980s pointed to the potential for creating a formal management process that could be applied by appropriately trained professionals across many disciplines. In 1990, Decision Partners was formed, comprising a small cadre of scientists, engineers, and management professionals. They established the Mental Modeling management process and began applying it to challenges worldwide. In the process, Decision Partners become the first commercial organization to harness the Mental Modeling approach to risk communication and put it to work addressing important and practical challenges.

Over the ensuing decades of application, Mental Modeling has been in continuous development, extending from initial applications in risk communication and risk management into a wide variety of other applications, from strategic planning, to stakeholder engagement to change management and technology transfer and adoption. Hundreds of applications of Mental Modeling, involving thousands of people worldwide, have been conducted by Decision Partners, academics at Carnegie Mellon and other universities, and other practitioners in industry and government, in particular colleagues at the U.S. Army Engineer Research and Development Center (ERDC), a branch of the U.S. Army Corps of Engineers (USACE). These many applications of Mental Modeling have evolved the practice of Mental Modeling and generated a continually expanding volume of knowledge products in the form of know-how and application experience.

This book taps that expertise and experience by offering an overview of the Mental Modeling methodology and providing a range of case examples of its application. In the chapters that follow, the authors describe the Mental Modeling management process in detail and provide a range of case study applications to help readers and risk communication practitioners to see how Mental Modeling can be applied to the challenges they face in better informing key decisions and shaping appropriate risk behavior. In the final chapter, the authors discuss new custom software developed by Decision Partners to support and enable practitioners to more effectively and efficiently apply Mental Modeling.

---

[1] Dr. Baruch Fischhoff is Howard Heinz University Professor with the Departments of Social and Decision science and of Engineering and Public Policy at Carnegie Mellon University, and Chief Scientist at Decision Partners.

## Understanding People's Mental Models

*Mental models* is a well-established concept in psychology and has been the focus of extensive research since the 1930s (Bartlett 1932; Johnson-Laird 1983; Bostrom et al. 1992; Atman et al. 1994; Fischhoff et al. 1997). Mental models have been described as representations or schemes of how people perceive and understand the world around them. A person's mental model can be thought of as a complex web of deeply and often subconsciously held beliefs that influence how an individual learns, defines a problem, reacts to information, forms judgments, and makes decisions. Mental models often define boundaries of thought and action, limiting people to familiar patterns of reasoning and action and tend to prevent them from seeing alternative perspectives.

## Mental Modeling: Critical to Effective Risk Communication

Decades of research and experience have shown that to effectively engage people through communications and enable changes in their beliefs and behaviors, one must first understand their mental models, then design strategies and communications to: reinforce what they know that is correct, address key knowledge gaps and misunderstandings, and reinforce judgments of credibility of the communications and their source.

The complexity of people's thinking makes it impossible to predict mental models with confidence or to predict the effects of information on people's mental models. Like all models, people's mental models are an abstraction of reality. They may be complete and correct, or they may have gaps or inconsistencies that are consequential to effective decision making and action. A mental model is usually less complex than the real-world phenomenon involved and tends to lag in context or time and so can easily become out of date. In many cases, people may lack conscious, well-formed mental models on issues that they have not thoroughly considered in the past. This may be challenging for decision-makers as people's responses may seem unpredictable or irrational.

Mental Modeling enables discovery of people's mental models in a structured, rigorous, respectful manner. Mental Modeling has been recognized as one of the premier methods for informing the development of strategies and communications that precisely address people's current thinking, judgment, decision making, and behavior on complex issues, including risk issues. Broadly, Mental Modeling works from the "inside out," starting with an in-depth understanding of people's mental models, and then using that insight to develop focused strategies and communication that builds on where people are at in their thinking today, reinforcing what they know about a topic and addressing critical gaps. Broadly stated, the goal is to help people make well-informed decisions and take appropriate actions on the topic at hand.

Effective analyses of mental models enable practitioners to identify gaps and alignments among the values, perceptions, decisions, and information needs of differing cohorts or stakeholder groups, identifying how they think about and respond to essentially any topic. It is particularly effective for complex challenges that are both socially and technically complicated.

The method relies on data collected from experts and nonexpert stakeholders (laypeople), typically through focused interviews with individuals conducted in person or by telephone. These insights are then coded, analyzed, and reported, enabling practitioners to develop policies, strategies, and communications that support well-informed decision making—by both the decision-makers and the stakeholders—that lead to appropriate behavioral outcomes.

The one-on-one mental models interviews lead participants through an agenda of topics in a context that approximates the natural decision-making environment, allowing for free expression and encouraging elaboration on topics in order to reveal individual perspectives at considerable depth. The open nature of the interviews allows interviewees to readily raise topics that they feel are important, even if they have not been anticipated in the research design and are not explicitly addressed in the agenda of topics.

The detailed, in-depth responses elicited from each interviewee allow for structured analyses, supported by software optimized for the Mental Modeling approach. When done well, analysts can identify key influences on people's judgment and decision-making processes, and their behavior. Analysts can also reveal key influencing beliefs and the underlying rationale for those beliefs. They are also able to compare analyses over time and provide insights into why beliefs may have changed.

## Key Benefits of the Mental Modeling

Mental Modeling is among the most robust of qualitative research methods. It yields rich, high-quality data on complex thinking, by intensive study of relatively small samples of strategically selected individuals (Morgan et al. 2002). Mental Modeling has been recognized within the U.S. Army Corps of Engineers (USACE); U.S. Department of Health and Human Services (HHS); the U.S. Food and Drug Administration (FDA; Fischhoff et al. 2011); and other U.S. federal agencies, Health Canada and the Public Health Agency of Canada, as well as many other organizations, for providing a solid foundation for evidence-based strategies and communications, as well as being a key part of an integrated risk management/risk communications approach (Health Canada 2006). The steps of the process can be tailored to best fit the challenge at hand, within the time, resource, and other constraints, and can be conducted to a high scientific standard. Further, new software support tools for Mental Modeling, described in Part IV, enable more effective and efficient knowledge integration, management, and communication.

A key strength of mental models research is its flexibility, allowing for remarkable integration with other research methods. Mental models research uncovers perceptions in depth, which can then provide critical insights for framing questions in surveys or developing discussion guidelines for deliberative dialogue or focus groups. In addition, when mental models research is conducted with experts, the resulting *expert model* (described in detail in Part I) can be used as an analytical framework for mental models research. It also enables integration of data generated through other forms of research including qualitative survey research and quantitative research, including risk assessment, and risk evaluation methods such as multi-criteria decision analysis and risk ranking (Belton and Stewart 2002; Keeney and Raiffa 1976; Linkov and Moberg 2011).

Finally, the process of conducting mental models research, a key component of Mental Modeling, treats stakeholders and their interests and priorities with respect, and is well suited to engaging stakeholders on sensitive risk topics often enhancing the perceived trustworthiness and competence of the sponsoring organization.

## Applied Mental Modeling

Decision Partners, USACE, the United States Food and Drug Administration (FDA) and many other prominent organizations have applied Mental Modeling to guide risk management and risk communications strategies and messages related to a diverse set of challenges, including sensitive public health and safety issues such as: drug safety and drug efficacy, food safety, plastic surgery, obesity, health impacts of extreme heat events resulting from climate change, and childhood vaccinations; science and technology issues such as: impacts of coal and nuclear power generation technologies; and environmental issues such as flood risk management, climate change, and environmental remediation.

In Part I, we describe the current Mental Modeling process as well as its historic foundations. This is done by first describing the current state of the process with a case study example from the health care industry, followed by a review of the process' historic development in the context of related modeling paradigms. In Part II, we discuss four USACE Mental Modeling case studies: Flood Risk Management, Beneficial Use of Sediment, Adaptive Management for Climate Change, and Technology Infusion and Marketing. In Part III, we offer a variety of case studies based on collaboration with a diverse group of clients, including: FDA's interest in discovering influences on Dairy Farmers' decision making to put dairy cows with residual pharmaceuticals into the food stream; the Chamber of Mines of South Africa's desire to significantly improve occupational health and safety performance in mining in South Africa; the nuclear industry's desire to engage stakeholders in a meaningful dialogue about the benefits and risks of their industry and its products and services, to name just a few of the case study applications in this book.

Since 2010 Decision Partners has been developing CASS™—Cognitive Analysis Software Suite—an integrated set of custom software tools designed specifically to efficiently support the essential Mental Modeling tasks of qualitative modeling and qualitative data analysis. CASS™ enables mental models research to be conducted more rapidly, accurately, and cost effectively than is possible manually. CASS™ for Mental Modeling is discussed in more detail in Part IV and a demonstration version of the software with sample models has been prepared especially for this publication and is available from Decision Partners.

## Who Should Use Mental Modeling?

The success of leaders in all organizations often turns on their ability to better understand and influence the judgment, decision making, and behavior of the people key to accomplishing their goals. Mental Modeling has proven to be well suited to this task over decades of application. Application experience demonstrates that, when supported with the relevant expertise, Mental Modeling can be successfully deployed by professionals in many disciplines including engineering, communications, and management.

This trend of expanding the professional practice of Mental Modeling is greatly expanding the total field of application worldwide. It is expected to accelerate significantly through the use of social media and the Internet, enabling unbounded collaboration among users of the methodology and expert application professionals. Fundamentally, Mental Modeling is about understanding the way that people think about issues and processes and using that insight to develop strategies and to manage the effects of communications, decision making, and behavior. CASS™ facilitates migration of Mental Modeling capabilities to a wider range of professionals than previously realized, creating a worldwide community of practitioners accelerating and extending Mental Modeling into new domains and increased globalization of its application.

## Overview of the Chapters

### *Part I: The Mental Modeling Approach*

In Part I, the authors provide an overview of the Mental Modeling process.

**Chapter 2: Mental Modeling Research Technical Approach** uses a case study in health decision making to describe and illustrate the key steps of the Mental Modeling process.

**Chapter 3: Science of Mental Modeling** gives intellectual context for the method, describing not only the historic roots of the process, but the relationship between Mental Modeling and other recent cognitive mapping techniques.

### *Part II: Applications in USACE*

In Part II, applications of Mental Modeling within USACE are provided.

**Chapter 4: Flood Risk Management** introduces Mental Modeling as a tool for process improvement in the context of the USACE flood risk management enterprise.

**Chapter 5: Adaptive Management for Climate Change** presents a case study that uses mental modeling to understand the relationships between climate change, current USACE risk management processes, and the actions that USACE and its partners can take to improve the nation's ability to adapt to climate change impacts.

**Chapter 6: Technology Infusion and Marketing** applies Mental Modeling techniques to identify opportunities for improving and sharing the USACE Environmental Laboratory's technical competencies with partners and potential clients.

## *Part III: Applications in Other Contexts and Industries*

In Part III, we illustrate the broad range of challenges that have been addressed with Mental Modeling by presenting several case studies in distinct domains.

**Chapter 7: Farmers' Decision Making to Avoid Drug Residues in Dairy Cows** demonstrates the application of Mental Modeling to understand the decision-making process that farmer's use when deciding whether and how to use medications, particularly antibiotics, with their livestock.

**Chapter 8: Influence of the CHEMM Tool on Planning, Preparedness, and Emergency Response: a Customized Strategic Communications Process Based on Mental Modeling** Mental Modeling as a foundation for developing a web portal (Chemical Hazards Emergency Medical Management Tool) to inform first responders, doctors, and citizens about the properties of common hazardous chemicals and the proper response if one comes into contact with those chemicals.

**Chapter 9: The Chamber of Mines of South Africa Leading Practice Adoption System** details the use of Mental Modeling for designing an industrial initiative to achieve systematic adoption of innovative technology and best practices.

**Chapter 10: Conducting Effective Outreach with Community Stakeholders about Biosolids** combines Strategic Risk Communications™ with Mental Modeling to better understand judgments of biosolids land applications across a wide range of stakeholders.

**Chapter 11: Using Mental Modeling to Systematically Build Community Support for New Coal Technologies for Electricity Generation** describes an application of Mental Modeling within a rural community in Alberta, Canada.

**Chapter 12: Saving Lives from a Silent Killer: Using Mental Modeling to Address Homeowners' Decision Making about Carbon Monoxide Poisoning** describes how Mental Modeling was used to assist the Technical Standards Safety Authority of Ontario and the Carbon Monoxide Safety Council in developing insight-based risk communications focused on raising homeowner awareness of the risk of CO in the home and the need to take appropriate action.

**Chapter 13: U.S. Census Bureau Integrated Communications Services for Data Dissemination: Mental Modeling Case Study with Key Internal Expert Stakeholders** presents an application of Mental Modeling applied to strategic stakeholder engagement to establish a customer-centric and data-driven method of communications to support the Census Bureau's mission to increase availability of data and survey analyses to the general public and other key audiences.

## *Part IV: Mental Modeling Software Support*

**Chapter 14: Mental Modeling Software Support** introduces CASS™ software, developed to support Mental Modeling.

## References

Atman, C. J., Bostrom, A., Fischhoff, B., & Morgan, M. G. (1994). Designing risk communications: Completing and correcting mental models of hazardous processes, part I. *Risk Analysis, 14*(5), 779–788.

Bartlett, F. C. (1932). *A theory of remembering* (pp. 197–214). London: Cambridge University Press.

Belton, V., & Stewart, T. J. (2002). *Multiple criteria decision analysis: An integrated approach*. New York: Springer.

Bostrom, A., Fischhoff, B., & Morgan, M. G. (1992). Characterizing mental models of hazardous processes: A methodology and an application to radon. *Journal of Social Issues, 48*(4), 85–100.

Fischhoff, B., et al. (1997). *Risk perception and communication* (Oxford textbook of public health, pp. 987–1002). London: Oxford University Press.

Fischhoff, B., Neuhauser, L., Paul, K., Brewer, N. T., & Downs, J. (2011). *Communicating risks and benefits: An evidence-based user's guide*. Silver Spring, MD: Food and Drug Administration.

Health Canada, Canadian Standards Organization (2006). A framework for strategic risk communications within the context of health Canada and the PHAC's integrated risk management, 0-662-44597-X. Retrieved January 28, 2016 from http://www.hc-sc.gc.ca/ahc-asc/pubs/_ris-comm/framework-cadre/index-eng.php.

Keeney, R. L., & Raiffa, H. (1976). *Decisions with multiple objectives: Preferences and value tradeoffs*. New York: Wiley.

Linkov, I., & Moberg, E. (2011). *Multi-criteria decision analysis: Environmental applications and case studies*. Boca Raton, FL: CRC Press.

Johnson-Laird, P. N. (1983). *Mental models: Towards a cognitive science of language, inference, and consciousness (No 6)*. Cambridge, MA: Harvard University Press.

Morgan, M. G., Fischhoff, B., Bostrom, A., & Atman, C. (2002). *Risk communication: A mental models approach*. New York: Cambridge University Press.

# Part I
# The Mental Modeling Approach

# Chapter 2
# Mental Modeling Research Technical Approach

**Sarah Thorne, Gordon Butte, Daniel Kovacs, and Matthew D. Wood**

## Introduction

The Mental Modeling research approach discussed in the following chapters is built on the foundational work in risk perception and risk communications at Carnegie Mellon University led by Dr. Baruch Fischhoff[1] and is well established in the fields of risk analysis and decision sciences (Atman et al. 1994; Bostrom et al. 1992; Fischhoff et al. 2011; Morgan et al. 2002). Mental Modeling is particularly well

This Guideline (subsequently revised in 2009 as Q850-87 (R2009) Risk Management: Guideline for Decision Makers) is also aligned with the US Presidential/Congressional Commission on Risk Assessment and Risk Management Process and the Australian/New Zealand Risk Management Standard. In addition, our work in strategic risk communications is aligned with the International Organization for Standardization's (ISO) 31000 Guidelines on Risk Management (2009), to which we provided input.

[1] Dr. Fischhoff is Decision Partners' Chief Scientist responsible for strategic research design, implementation, and analysis. He is also the Howard Heinz University Professor of the Departments of Social and Decision Science, and Engineering and Public Policy at Carnegie Mellon University.

S. Thorne, M.A.
Decision Partners, 1084 Queen Street West, #32B, Mississauga, ON, Canada L5H 4K4
e-mail: sthorne@decisionpartners.com

G. Butte (✉)
Decision Partners LLC, Suite 200, 313 East Carson Street, Pittsburgh, PA 15217, USA
e-mail: gbutte@decisionpartners.com

D. Kovacs, Ph.D.
Decision Partners, 1458 Jersey Street, Lake Milton, OH 44429, USA
e-mail: dkovacs@decisionpartners.com

M.D. Wood, Ph.D.
U.S. Army Corps of Engineers, Engineer Research and Development Center (ERDC) and Carnegie Mellon University, 696 Virginia Road, Concord, MA 01742, USA
e-mail: Matthew.d.wood@usace.army.mil



M.D. Wood et al., *Mental Modeling Approach*, Risk, Systems and Decisions,
DOI 10.1007/978-1-4939-6616-5_2

suited for generating the in-depth, evidence-based understanding of factors influencing decision making and behavior required to develop strategies, plans, and communications to effectively address people's thinking on complex issues. The process is *science-informed*, based on social science methodology, and *evidence-based* in that it facilitates the use of information systematically gathered from stakeholders themselves. Its purpose is to help decision-makers and communicators make informed decisions about how best to communicate risks, design policy, or develop behavioral interventions with the needs, priorities, and interests of the focal stakeholders in mind.

The central idea behind Mental Modeling is that people's judgments, decision making, and behavior about whether and how to adopt a new innovation, accept a medical procedure, or support a power plant or natural gas transmission line, are influenced by their mental models (Morgan et al. 2002).

*Mental models* are the tacit webs of beliefs that individuals draw upon to interpret and make inferences about issues that come to their attention. They develop over time based on a person's values, priorities, experiences and observations, formal education, and communications of all kinds. Where persons have no experience upon which to draw, they will draw inferences from existing mental models that seem relevant to them (Fischhoff et al. 2002). Information perceived as consistent with existing beliefs is readily incorporated into a person's mental model; information at odds with existing beliefs is not, and may even be rejected.

## Overview of Mental Modeling Research Methodology

The concept of mental models has been the focus of extensive research in the field of psychology dating back to the 1930s. A person's mental model can be thought of as a complex web of deeply held beliefs below the surface of conscious thinking that affect how an individual defines a problem, reacts to information, forms judgments, and makes decisions. One's beliefs about a topic may be complete and correct, or they may have consequential gaps and misperceptions that negatively influence decision making and action—behavior. Mental models are not observable; they can only be determined with empirical research. They are typically represented using influence diagrams which depict the factors a person perceives as relevant to the issue at hand, with directional arrows showing how the value (or level) of one factor influences the value of another (Johnson-Laird 1983).

Decades of research and experience have shown that to effectively engage people through communications and enable changes in their beliefs and behaviors, one must first understand their mental models. Once these models are understood, one can then design strategies and communications that: reinforce what they know that is correct, address key knowledge gaps and misperceptions that are consequential, and use communications sources and methods that are credible and relevant to the focal stakeholders. Research into individuals' mental models reveals critical issues and identifies gaps and alignments among the values, perceptions, decisions, and information needs of the various stakeholder groups.

An *expert model* is an important element in Mental Modeling. It is a formal, comprehensive graphic representation that summarizes and integrates the current knowledge and understanding of experts about the key factors of the topic being studied. It can be thought of as an "expert's mental model," as it typically comprises a composite of the knowledge and beliefs—mental models—of several experts. Expertise is often distributed throughout the stakeholder community and may be formal or informal. For complex situations or problems, an expert model captures the breadth of expertise that is often distributed across a number of *experts*, each with specific areas of expertise. As a depiction of experts' understanding of a topic, or their contribution to the topic, an expert model is expected to be relatively accurate and objective if the experts participating in the model development have the requisite expertise to address the major factors in the model being depicted. That said, as described in more detail later, the Mental Modeling approach is specifically designed to reveal the factors that stakeholders believe to be relevant even if those factors have not been anticipated by the research designers or participating experts. Often stakeholder interviews reveal factors that the experts have not considered. Such discovery is a benefit of the Mental Modeling approach that cannot be replicated with opinion surveys or other tools designed to assess how many people think the same thing about a set of prescribed factors.

Expert models are essential management tools used to ensure that a project team and key stakeholders are aligned on the understanding of the topic at hand and the project scope. They also serve as the analytical framework for the design, implementation, and structured analyses of mental models research. The focus of such research is to provide deep insight into nonexpert stakeholders' (laypeople's) mental models of the topic at hand.

## Key Benefits of Mental Modeling

Mental Modeling is among the most robust of qualitative research methods. It yields rich, high-quality data on individuals thinking on complex topics, by intensive study of relatively small samples of strategically selected individuals (Morgan et al. 2002). Mental Modeling has been recognized within the U.S. Army Corps of Engineers (USACE), U.S. Department of Health and Human Services (HHS), the U.S. Food and Drug Administration (FDA; Fischhoff et al. 2011), along with many other federal agencies and other organizations, as providing a solid foundation for science-informed, evidence-based strategies and communications, as well as being a key part of an integrated risk management/risk communications approach (Standards Council of Canada 1997; ISO 2009).

Using an expert model as the analytical framework enables integration of data generated through other types of qualitative and quantitative research methods, such as focus groups, and surveys. The results of mental models research can be used in risk assessment and risk evaluation methods such as multicriteria decision analysis and risk ranking.

The process of conducting mental models research treats stakeholder interests and priorities with respect and is well suited to engaging stakeholders on sensitive topics. In addition, the act of conducting the research often enhances the perceived trustworthiness and competence of the sponsoring organization.

Over the past 30 years mental models research has been applied to guide risk management and risk communications strategies and messages related to a diverse set of challenges. These include: sensitive public health and safety issues such as drug safety and drug efficacy, food safety, plastic surgery, obesity, health impacts of extreme heat events resulting from climate change, and childhood vaccinations; science and technology issues such as impacts of coal and nuclear power generation technologies; and environmental issues such as flood risk management, climate change, and environmental remediation.

The varied and complex needs of Decision Partners clientele has led to advances on the approach described in Morgan et al. (2002) through wide adaptation and customization. This can be observed through the Strategic Risk Communications Process™, which aligns with the Canadian Standards Q850-97 Risk Management: Guideline for Decision-Makers, and was adapted for Health Canada and the Public Health Agency of Canada in 2006 as the base of their Strategic Risk Communications Framework and Handbook (Health Canada 2006). It was further customized and tested as part of a 2008–2011 research challenge supported by the U.S. Water Environment Research Foundation (WERF) and adapted for use by WERF and its members (Chap. 10; Eggers et al. 2011; Beddow 2011). Decision Partners has also developed software support tools for Mental Modeling (Cognitive Science Systems 2012) that can be customized to enable more effective and efficient knowledge integration, management, and communication. Mental Modeling Technology™ is a unique, evidence-based, science-informed management process for developing programs, i.e., policies, strategies, and communications, for belief and behavior change. The technology comprises integrated methods and tools on a software platform. The applied cognitive behavioral approach enables systematic formulation of strategies and communications for shaping judgment, decision making, and behavior.[2]

## Mental Modeling Core Technique

Mental Modeling Core Technique provides a basic illustration of the essential Mental Modeling approach, summarized in three phases (Fig. 2.1). The detailed 6-step process is described using a case study, beginning on the next page.

The Mental Modeling approach starts with developing a picture of the basic system related to the topic being studied. This model, called the Base Expert Model

[2] In February 2016, Decision Partners received a patent for its Mental Modeling Method. This patent reflects the essential intellectual property and software tools that comprise Mental Modeling Technology™.

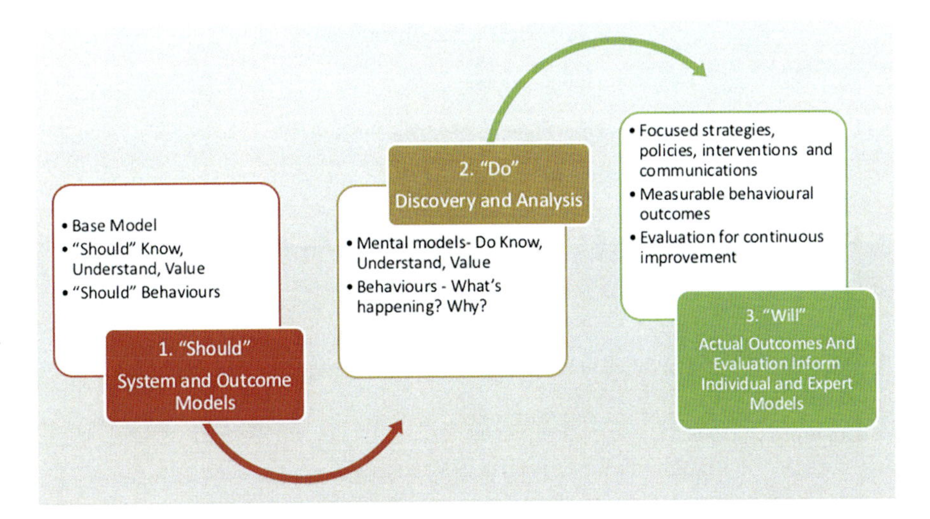


**Fig. 2.1** Mental modeling core technique

and described in detail in the case study that follows, is developed based on the subject matter knowledge and experience of the experts who are often members of the client project team, along with other key internal and external experts. The resulting expert model provides a system picture of what the experts believe the stakeholders *should* know about the topic and what behaviors they *should* take as a result.

In the next phase, mental models research is conducted with the key stakeholders to discover what they *will do* based on their mental models of the topic, which can be characterized as what they know that is correct, what they misunderstand, what they want to know, and who they trust and what communications processes they trust. This information, collected through one-on-one interviews, is coded and analyzed against the expert model, revealing alignments and gaps between the experts' and stakeholders' mental models.

In the final phase, the insight from the mental models research is used to develop focused strategies, policies, interventions, and communications to enable stakeholders to make well-informed decisions and take appropriate actions on the topic at hand, in short, to guide what they *will* do. It also involves measurement and evaluation of the strategies, policies, interventions, and/or communications to ensure that they are producing the desired behavioral outcomes.

## Key Steps in the Mental Modeling Process

The following are the key steps in the Mental Modeling process (Fig. 2.2). Each project is unique and not all steps are performed for every project. For example, for some policy or strategy challenges, an expert model is developed as the foundation. Sometimes Step 4 is conducted after Step 5. Some projects may follow an iterative,

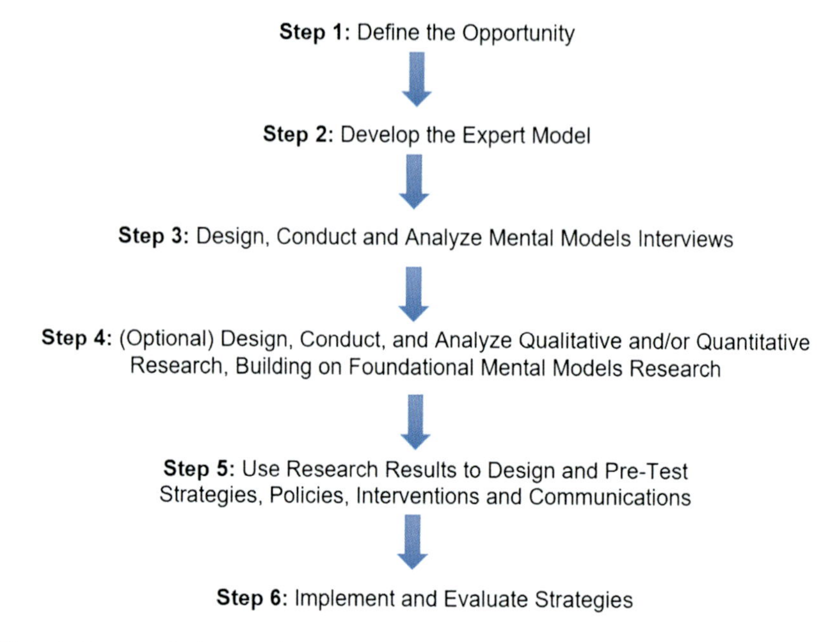


**Fig. 2.2** Key steps in the mental modeling process

multistage design, for example, where many rounds of stakeholder research are conducted. In some of the case studies that follow, the steps have been refined to reflect the focus of the project.

These steps are described later and are illustrated using content from a project conducted for the American Society of Plastic Surgeons (ASPS) to design a comprehensive communications strategy based on deep insight into potential patients' and consumers' mental models of plastic surgery and plastic surgeons. The mental models research phase was conducted in 2004 to define the strategy and content of the subsequent marketing campaign. The research was completed in 2005.

## Step 1: Define the Opportunity

Mental Modeling begins by preparing an Opportunity Statement, which clearly defines the goals and desired outcomes of the project, including the desired behavioral outcomes resulting from the application of the research. This statement frames and focuses the mental models research and its end results and ensures the project team is aligned in its understanding of the project's scope and focus, and that team members understand their respective roles.

### *ASPS Opportunity Statement Example*

The opportunity is to support the design and execution of a strategic, science-based campaign to encourage people considering plastic surgery to choose a Board Certified Plastic Surgeon. (Note: this project was conducted by Reingold,[3] Decision Partners and Penn Schoen Berland (PSB).[4] Here we focus primarily on the role of the mental models research in shaping the design of the campaign developed by Reingold and evaluated by PSB.) This will be achieved by:

- Characterizing expert knowledge of influences on decision making of prospective patients/customers, including perceived risks and benefits.
- Gaining in-depth insight into factors influencing prospective patient/customer decision making concerning having plastic surgery, including their perceptions and weighing of benefits and risks, their selection of a plastic surgeon and their judgment of the outcomes, along with understanding what information sources and communication methods are used and valued by potential patients/customers.
- Creating campaign strategies and messages addressing critical decisions prospective patients/customers face.
- Evaluating strategies and messages through empirical testing before deployment.
- Evaluating results, including communication design, messages, and channels, along with outcomes of strategies.
- Continuously improving by identifying and implementing improvements in communication processes based on evaluations.

## Step 2: Develop the Expert Model

As described earlier, an expert model summarizes and integrates expert knowledge on the topic of interest, typically using graphic decision-modeling representations, which provides an analytical framework for the design and analysis of later in-depth mental models research with key stakeholders. This framework facilitates later direct comparison between experts' and stakeholders' mental models of the topic. The model-creation process usually starts with a review of literature or relevant materials provided by the client, followed by informal, but in-depth, one-on-one interviews with a small number of experts or a workshop with a group of experts. Notes, recordings, or transcripts of the expert sessions are used as needed to support development of the expert model, though they are not typically formally coded and analyzed.

---

[3] Reingold is a small, full service communications firm based in Alexandra, VA.

[4] Penn Schoen Berland is one of the world's premier strategic opinion research and communications consulting firms, and is based in Washington, DC.

A common form of an expert model is an influence diagram, which represents knowledge in terms of variables and the relationships among them, as they relate to the outcomes of interest. The key influences or variables in the system are depicted as nodes, typically ovals, with descriptive titles and subtext and can depict a number of different types of influences, including the potential outcomes associated with the topic, the people whose decisions and behavior influence those outcomes, factors that "drive" or provide the basic context of the situation being studied, and the factors that influence these stakeholders' decision making and behavior.

The nodes are positioned to create a logical flow, typically starting with driving or foundational contextual variables in the upper left corner and ultimate decisions or desired outcomes positioned in the lower right corner of the model. Relationships and primary direction of influence among the variables are illustrated with arrows with the node at the tail of an arrow "influencing" the variable at the arrow's head. Additional graphic design concepts can be employed such as using color to semantically group similar nodes. This is particularly helpful in enabling users to "follow" nodes through various models that may present the variables at different levels of detail. Other techniques to represent expert knowledge may include logic models, decision trees, fault trees, and multiattribute matrices.

Expert models are created iteratively and final models are often validated by reviewing them with the experts who participated in their creation. Expert models can also be refined over the duration of a project to represent the most up-to-date knowledge on the topic as it evolves. Expert models provide the analytical framework for the design and analysis of in-depth stakeholder interviews, allowing direct comparison between experts' and stakeholders' mental models of the topic.

Expert models are typically not used as primary communications tools; however, they do provide a useful foundation upon which to engage experts and often key stakeholders in dialogue about the initiative. The system perspective enables all participants to understand all of the influences on the decision making of the focal stakeholders and to come to a shared understanding of the system drivers and desired outcomes.

## ASPS Expert Model Example

The *Base Expert Model of Influences on Potential Patient Decision Making Regarding Plastic Surgery* (Fig. 2.3) provides an overview of the context in which the decision to have plastic surgery is made by the consumer, as estimated by experts. Thirteen experts in cosmetic and reconstructive surgery were interviewed, and their knowledge was integrated to create an expert model illustrating their understanding of the key influences on individuals' decisions to have plastic surgery. Note the arrows or *influences* that link related *nodes* or variables that define the decision-making context. The experts believed that an individual's *Evaluation of Surgeon*, for example, would be a driving influence on their decisions. This Model

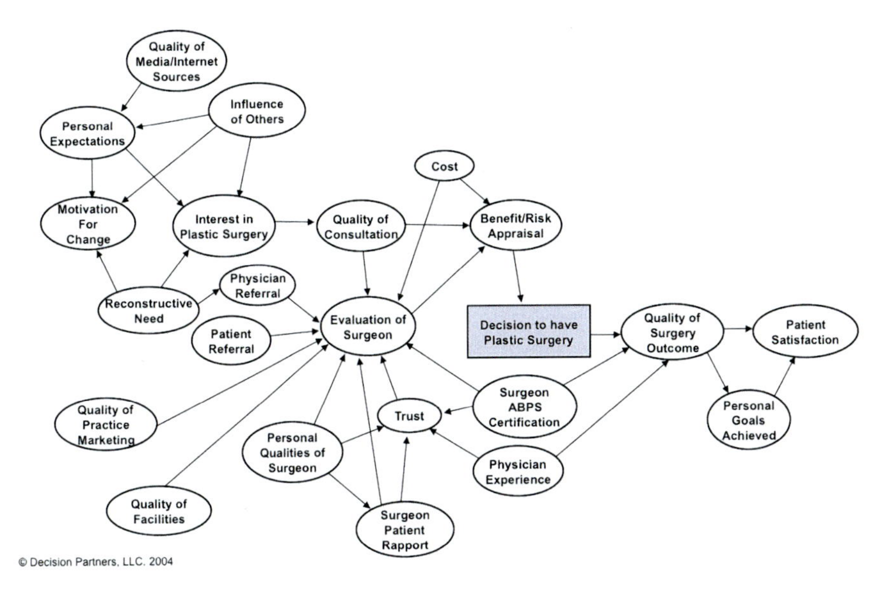


**Fig. 2.3** Base expert model of influences on potential patient decision making regarding plastic surgery

was then used to design the protocol for in-depth mental models interviews with potential patients and subsequent interview analysis.

The following narrative describes the Model.

## *Drivers*

The prospective patient's (or prospective customer's) decision to have plastic surgery involves a dynamic mix of external influences and individually intrinsic motivations. Media, family, peers, lifestyle, and stage of life are believed to all influence a potential patient's understanding and perception of plastic surgery as a desired option for self-change.

Motivation for self-change is thought to develop from a sense that one's physical, emotional, or social well-being could be improved through plastic surgery. Experts believe there is a relationship between expectations for and appraisal of one's physical well-being and one's feelings and relationships with others. Physical health and attractiveness are related to feelings of self-confidence and self-esteem.

Once motivated toward having plastic surgery, potential patients are believed to begin a process of research and consultation about the procedure(s) of interest and about available plastic surgeons. The depth and breadth of this research is thought to vary across individuals. Potential patients are believed to invest time speaking with plastic surgeons, other doctors, family and friends, and past patients. Referrals, consultations, and research about a surgeon's history and experience are believed to significantly influence patient selection of an individual plastic surgeon.

Practice marketing, such as print materials and surgeon websites, was thought to inform this research. The surgeon selection experience was considered to be an important part of the decision to have plastic surgery, and can be influenced by personal qualities of the plastic surgeon, as well as the degree to which he/she establishes rapport with the prospective patient and earns his or her trust.

Experts also thought potential patients' evaluation of a plastic surgeon is influenced by the quality of a surgeon's office, staff, and facilities. Additionally, potential patients considered surgeon experience, training and professional association, such as membership in the ASPS, in their selection of a plastic surgeon.

In making the decision to have plastic surgery, that is, an appraisal of risks and benefits associated with the procedure and surgery in general, experts said potential patients considered the associated costs, the plastic surgeon, and their confidence that their expectations will be met. At this point, potential patients may decide to have plastic surgery and may select a Board Certified plastic surgeon, or they may decide to pursue another option.

### *Outcomes*

Here, the "quality of the surgery outcome" refers to the patient's personal evaluation of his or her experience. The surgeon's experience, training, and professional memberships were thought to be influences on quality. As well, pre- and postoperative care and the occurrence of complications or unexpected outcomes play a role in the quality of outcome. Patient satisfaction is related to the perception that the outcome has met individual expectations and goals. Positive experiences and satisfaction may then lead to future benefits for the plastic surgeon, such as repeat patients and patient referrals.

## Step 3: Design, Conduct, and Analyze Mental Models Interviews

The next step in the Mental Modeling approach is designing and conducting one-on-one in-depth interviews following a semistructured interview protocol. The research sample of individuals representing the stakeholder population(s) of interest (or cohort) is the core of the Mental Modeling research approach. This sample is usually comprised of 20–30 individuals, each representing a focal stakeholder. This stratified sampling is done in order to reveal the breadth of perceptions held. Research interviewees are selected from a larger pool of individuals to allow for random sampling and to provide a level of confidentiality. Subcohorts may also be used, or a matrix cohort design may be used to ensure representation of gender or other demographic factors.

Mental Models interviews follow a semistructured interview protocol designed to explore key topics identified in the expert model. Interviewers trained in the Mental Modeling approach are oriented to the project and trained on the interview protocol. Once the sample is developed, the interviews are conducted, typically over the phone, but sometimes in person if appropriate or required.[5] Interviews are recorded with interviewee's permission and transcripts are produced and used as the primary data in structured analyses.

Questions, particularly early in the interview, are typically structured to elicit people's mental models using a "what comes to mind when you think about approach," asking the interviewee to think freely about a general topic rather than respond to a more narrowly focused question. Interviewers will also use general prompts such as, "Can you tell me more about that?" or "Why do you say that?" to probe interviewee responses, encouraging them to speak at length. This approach is specifically designed to allow topics of interest to the interviewee to more readily emerge, using the language and terminology that they would normally use. As the interviews progress, more specific and directed questions will be used to ensure coverage of all relevant variables in the expert model.

The interview data are then coded and analyzed against the expert model in order to describe stakeholders' beliefs about the topic including: their values, interests, and priorities; what they know; what they don't know or misunderstand; what they want to know; and who and what communications processes they trust. Depending on the needs and complexity of the project, formal or informal coding approaches can be applied. For less complex projects where one simply needs to summarize the prevalence of perceptions and beliefs, a basic, one-pass direct coding process may be used linking interviewee responses to specific concepts.

For more complex challenges, where stakeholders' perceptions are likely to cover a broad spectrum of beliefs that are often more nuanced, or for projects that require application of more rigorous academic research standards for coding and analysis, a multiple-pass approach may be more appropriate. In a multiple-pass coding approach interviewee responses are first "tagged" to link responses to general topics (often expert model nodes or basic themes). This facilitates a more thorough exploration of the interview data than a linear, "by-question" coding process. In the second coding pass, responses are coded against more specific emerging themes. The prevalence of these themes is then enumerated and reported.

---

[5] In-person interviews can add considerable time and cost and may increase the potential for "please-the-interviewer" bias compared to phone interviews, which may be perceived as more equitable by participants.

The comparison of structured qualitative analysis of the interview results against the expert model enables identification of key areas of alignment and critical gaps between the expert knowledge and the thinking of stakeholders, identifying: what stakeholders know, what they don't know or misunderstand, what they want to know, and who and what communications sources and methods they trust. This analysis provides the requisite insight to develop precisely targeted strategies, policies, interventions, and/or communications materials with clear, measurable behavioral outcomes.

## *ASPS Protocol Example*

**Excerpt from ASPS Interview Protocol** Our discussion will focus on your expectations and considerations about having plastic surgery. Later in the discussion we will discuss what is important to you when selecting a physician to perform the procedure(s). We'll conclude by talking about communications associated with plastic surgery. But again, I want to make sure I hear everything you have to say so if other topics or thoughts come to mind, please share them with me.

**Section 1: Motivation for Plastic Surgery Option. Perceived Benefits and Risks**

There are lots of reasons why people have plastic surgery. I'd like to hear about your interest in having plastic surgery.

1. Do you recall what first got you interested in having plastic surgery?
   - *(Probe as appropriate)* Can you tell me more about what interested you about plastic surgery at that time?
   - When was that?
2. In what ways do you hope to benefit from plastic surgery?
   - *(Probe for the following if not already mentioned)* What do you see as the benefits of plastic surgery to your physical appearance?
   - What do you see as benefits when it comes to how you feel about yourself?
   - What about when it comes to how others may feel about you and treat you?
   - Might there be any economic benefits as a result of having the plastic surgery procedure(s) you are considering?

## *ASPS Top Line Findings*

The following textbox presents an example of select top-line findings of research conducted with 60 people who had either undergone plastic surgery within the past 18 months (37 % of the sample) or who were actively considering cosmetic surgery (63 %). One-on-one confidential interviews, averaging 35 min, were conducted by phone.

**Select Top-Line Findings of People Who Had or Were Actively Contemplating Plastic Surgery**

The research findings were contrary to many widely held beliefs and conventional wisdom about who had plastic surgery and why they had it. Results included the following:

- Interviewees were motivated to have plastic surgery because they were unhappy with some physical aspect of their bodies. Many had been thinking about it for a long time, some their entire adult lives.
- The journey for most, from initial interest to undergoing an actual plastic surgery procedure, took months or years. Interviewees started by actively seeking information from others who had undergone similar procedures, then proceeded to conducting Internet research on the specific procedure they were interested in, followed by conducting research on and consulting with two or three individual surgeons.
- The primary benefit interviewees anticipated was feeling better about the way they look, which they believed would result in feeling more confident. Most said they had or would have plastic surgery to suit themselves, not the interests of others.
- Few interviewees spontaneously mentioned risks. When prompted, the most critical risk mentioned was not looking the way they expected as a result of the procedure. Nearly all believed the benefits of plastic surgery results far outweighed the risks.
- Interviewees' trust in their surgeon was a critical influence on their ultimate decision to have plastic surgery. Potential patients believed they could minimize the risks by "doing their homework" and selecting "the right" plastic surgeon.

Figure 2.4 shows how the interview results were used to develop weighted mental models diagrams, graphically depicting the influences on interviewees' decision making. The shading of the Mental Models diagram reflects that importance: "Primary influences" (red) were raised by more than 60 % of interviewees; "secondary influences" (orange) were raised by 20–60 %; and "other influences" (yellow) were raised by fewer than 20 % of the interviewees.

## Step 4 (Optional): Design, Conduct, and Analyze Qualitative and/or Quantitative Research, Building on Foundational Mental Models Research

Where needed, one can follow-up, supplement, or complement in-depth qualitative (mental models) interviews with other qualitative research such as focus groups, or quantitative research in the form of structured surveys with a large and representative

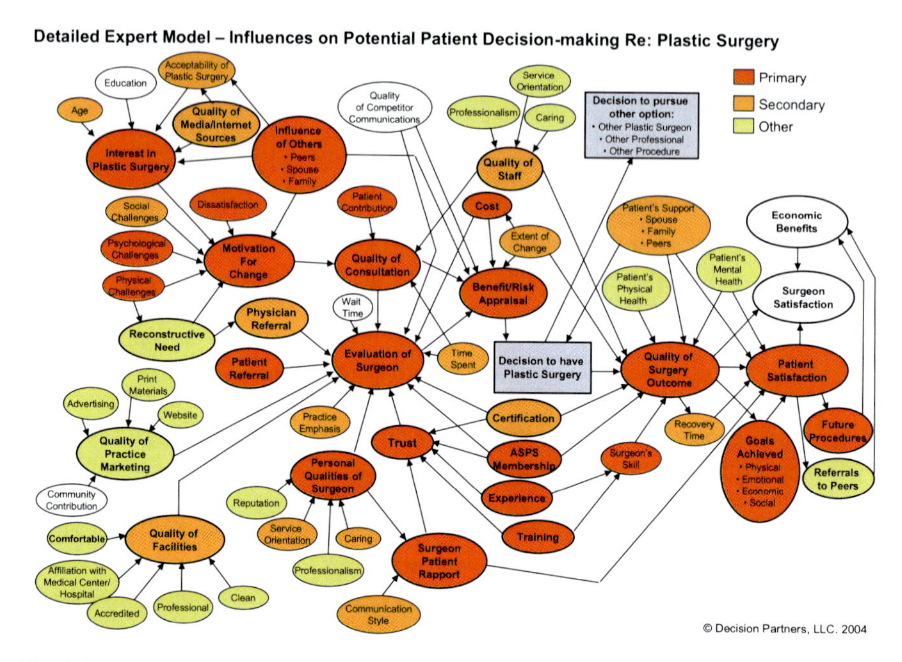


**Fig. 2.4** Sample mental models diagram

sample of stakeholders, or other quantitative methods. Such research can be an efficient and economical approach to validate and extend the results of mental models interviews in differing social contexts and, in the case of representative surveys, can be used to quantitatively assess the prevalence of beliefs in that population.

Figure 2.5 presents sample results from a follow-on, quantitative web survey[6] with U.S. adults who said they were considering plastic surgery or another appearance-altering procedure in the next year or two. The mental models research protocol and results were used to design the web survey questionnaire. The open-ended Mental Modeling approach allowed for characterizing beliefs and influences relevant to the decision to have plastic surgery. The green bars in Fig. 2.5 show the relative importance of different considerations in these decisions regarding plastic surgery (which 19 % had undergone before) for the survey sample. The black bars show comparable ratings to the mental models interviews. With few exceptions, the results from the web survey and the interviews were very similar.

## Step 5: Use Research Results to Design and Pretest Strategies, Policies, Interventions, and Communications

As described earlier, the mental models research results identify key areas of alignment and critical gaps between expert and nonexpert stakeholders. With this knowledge, precisely targeted strategies, policies, interventions, and communications

[6] Conducted by Penn Schoen Berland.

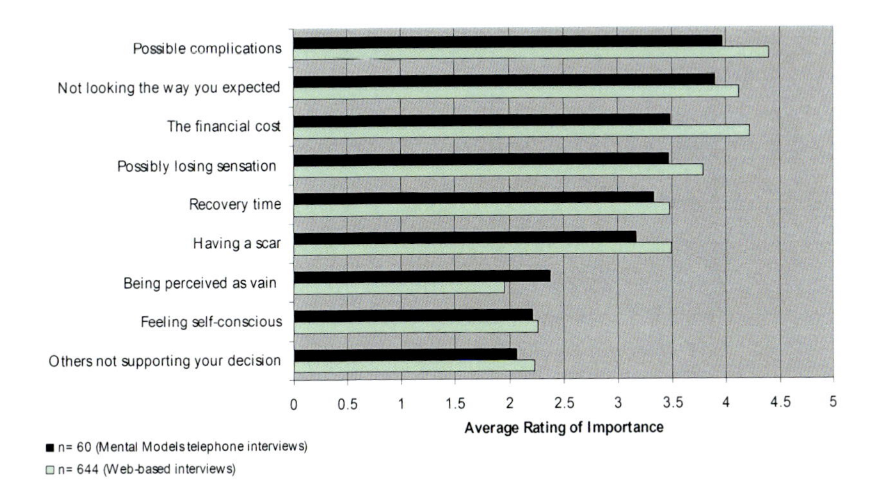


**Fig. 2.5** Interview and survey participants' ratings of potential risks of plastic surgery

plans and messages can be designed to address the critical influences on stakeholders' decision making and, ultimately, behavior related to the topic at hand. Communications should be designed to: reinforce what people know that is correct, close critical knowledge gaps and correct misperceptions, and address specific questions that people expressed during the research. Strategies, plans, and messages should be designed to enable stakeholders—including the decision-makers—to make appropriate decisions and judgments and, where appropriate, take action. In some cases, an intervention is required immediately to address a potentially risky situation or behavior. Consequently, some or all of the desired outcomes are behavioral, that is, focused on measurable actions. Strategies should be designed to use sources and modes of communications in which stakeholders have expressed a preference, and in which they have expressed a level of trust and confidence.

Messages that are developed should be pretested before being implemented more broadly, to ensure they have the intended effect.[7] A number of techniques can be used to test messages and materials, from small-scale read backs, to message-focused mental models research, to self-administered surveys. Methods can incorporate online components and visual testing. Choosing an appropriate technique depends on the nature of the materials, the stakeholder or audience for whom it is intended, and the amount of time and resources available. There is no formula for selecting a pretesting technique, nor is there a perfect technique for pretesting. The method should be selected and shaped to fit the pretesting requirement and the time and resources available.

[7] Such testing can also be conducted to evaluate performance of current or past strategies and communications for purposes of identifying improvements to both.

### *ASPS Strategy Example*

Building on the results of the mental models research and supporting web survey results, the challenge was to develop a brand and identity for ASPS that differentiated its members from the many other providers of plastic surgery procedures. At the same time, the brand and identity would need to focus on patient safety, how to choose an appropriate plastic surgery provider, and promote and advance the image of plastic surgery by highlighting the good work of ASPS surgeons. Focused benefit/risk communications plans, messages, and materials were developed to support plastic surgeons and prospective patients; shape industry association policies, procedures, and training related to conducting appropriate risk communications and informed consent dialogue with patients; and broaden industry outreach and dialogue with key stakeholders, including policy makers.

Key goals for the research-informed campaign included:

- Branding ASPS member surgeons;
- Differentiating ASPS member surgeons from other providers performing plastic surgery procedures, with emphasis on their training and board certification;
- Elevating the image of plastic surgeons among other medical specialists and the general public; and
- Educating the public and prospective patients about safe plastic surgery and the need to choose a Board Certified plastic surgeon.

Subsequent message testing research revealed that people considering plastic surgery and the general public were not knowledgeable about who could perform plastic surgery or what differentiated the qualifications of an ASPS board certified plastic surgeon. Communicating the key differentiator—5 years of surgical training with two additional years in plastic surgery—was key to the campaign and critical to helping prospective patients make well-informed decisions. An important component of this was providing tools to support prospective patients in "doing their homework" (as they called it in the mental models research) to select the best plastic surgeon most suited to working with them.

## Step 6: Implement and Evaluate Strategies

Throughout implementation, strategies and communications messages and delivery channels are adapted, enhanced, and modified as necessary. For projects that extend over many years and/or require sustainable behavioral outcomes, frequent evaluation and refinement are critical to success.

## *ASPS Case: Implementation Results*

A comprehensive campaign comprising advertising, video blogs, partner outreach, discussion forums, and decision support tools was developed, pretested, implemented, evaluated, and refined over several years. One example of the outreach materials developed to support patient engagement and decision making—the Find-a-Surgeon Tool—was improved based on testing and user feedback, is illustrated in Fig. 2.6.

Fig. 2.6 Example of outreach material—find a surgeon tool

In summary, the research-driven ASPS campaign was judged by the client to be very successful in achieving all its goals. Key measurements included the following:

- 350,000 online referrals per year;
- 46 % increase in awareness and value of the ASPS Member Surgeon brand; and
- 92 % increase in online engagement of potential clients of plastic surgery.

**Acknowledgment** Special thanks to Tanya Darisi, Robert Green and Joseph Ney for their contributions to the chapter.

## References

Atman, C. J., Bostrom, A., Fischhoff, B., & Morgan, M. G. (1994). Designing risk communications: completing and correcting mental models of hazardous processes, part I. *Risk Analysis, 14*(5), 779–788.

Beddow, V. (2011). *Conducting effective community outreach and dialogue on biosolids land application: Primer for biosolids professionals*. Alexandria, VA: IWA Publishing.

Bostrom, A., Fischhoff, B., & Morgan, M. G. (1992). Characterizing mental models of hazardous processes: A methodology and an application to radon. *Journal of Social Issues, 48*(4), 85–100.

Cognitive Science Systems, LP. (2012). ECASS: Expert modeling module of the cognitive analysis software suite (computer software). Retrieved May 10, 2012, from http://www.decisionpartners.com.

Eggers, S., Thorne, S., Butte, G., & Sousa, K. (2011). *A strategic risk communications process for outreach and dialogue on biosolids land application*. Alexandria, VA: Water Environment Research Foundation.

Fischhoff, B., Brewer, N. T., & Downs, J. S. (2011). *Communicating risks and benefits: An evidence-based user's guide*. Silver Spring, MD: US Food and Drug Administration.

Health Canada, Canadian Standards Organization. (2006). A framework for strategic risk communications within the context of health Canada and the PHAC's integrated risk management, 0-662-44597-X. Retrieved January 28, 2016 from http://www.hc-sc.gc.ca/ahc-asc/pubs/_riscomm/framework-cadre/index-eng.php.

ISO. (2009). *31000: 2009 Risk management–principles and guidelines*. Geneva, Switzerland: International Organization for Standardization.

Johnson-Laird, P. (1983). *Mental models*. Cambridge: Harvard University Press.

Morgan, M. G., Fischhoff, B., Bostrom, A., & Atman, C. (2002). *Risk communication: A mental models approach*. New York: Cambridge University Press.

Standards Council of Canada. (1997). *Risk management: Guideline for decision-makers*. Etobicoke, ON, Canada: The Canadian Standards Association.

# Chapter 3
# Science of Mental Modeling

**Matthew D. Wood and Igor Linkov**

## Mental Modeling as Evidence-Based Practice

Mental Modeling aggregates information about a problem or system in a deliberate and structured way to facilitate evidence-based practice in a wide host of domains. Current risk management practice uses conceptual models of important components of a problem and their relationships in an ad hoc manner (Kiker et al. 2005). Stakeholder perceptions and beliefs, including those of the general public, are not usually visualized in a structured way in current practice. Mental Modeling, in contrast, provides a systematic way to collect information in service of achieving some desired end state. This can be a piece of risk communication, public policy, engineered infrastructure, etc. Despite the increasing importance of public input in government, corporate, and organizational decision-making processes, it is challenging for decision-makers to identify and incorporate diverse perspectives and beliefs into risk management. Mental Modeling provides a tool to help overcome this challenge.

---

This chapter is adapted from work by Wood et al. (2012). Cognitive mapping tools: Review and risk management needs. Risk Analysis, 32(8), 1333-1348. Special thanks to Dr. Ann Bostrom, University of Washington, and Dr. Todd Bridges, U.S. Army Corps of Engineers, Engineer Research and Development Center, for their contributions to this chapter.

M.D. Wood, Ph.D. (✉)
U.S. Army Corps of Engineers, Engineer Research and Development Center (ERDC) and Carnegie Mellon University, 696 Virginia Road, Concord, MA 01742, USA
e-mail: Matthew.d.wood@usace.army.mil

I. Linkov, Ph.D.
U.S. Army Corps of Engineers, Engineer Research and Development Center (ERDC) and Carnegie Mellon University, 696 Virginia Road, Concord, MA 01742, USA
e-mail: Igor.Linkov@usace.army.mil



M.D. Wood et al., *Mental Modeling Approach*, Risk, Systems and Decisions,
DOI 10.1007/978-1-4939-6616-5_3

Mental Modeling as a methodology has proven effective as a robust qualitative research technique that yields rich data on the complex cognitive frameworks that individuals use when approaching a whole host of domains. Several such domains are identified in the case studies to follow in this chapter. The approach has been used by a variety of federal agencies as a firm foundation for evidence-based strategies. The approach can be modified to best address research needs while considering time, capabilities, and/or other resource and development constraints (Downs and Fischhoff 2011). The Expert Model is the foundation of the analytical framework that can facilitate data integration through additional quantitative or qualitative research, including risk assessment, risk ranking, and multicriteria decision analysis (Linkov et al. 2012). The approach has been applied to a broad range of challenges over the past 30 years, including the design and implementation of risk communications strategies and messages on topics as diverse as: drug safety and drug efficacy, food safety, plastic surgery, obesity, and childhood vaccinations (Downs et al. 2008; Fischhoff 1995), among others.

A brief background and synopsis of the Mental Modeling approach follows, with a summary and comparison to similar techniques designed to visually represent how people reason about a topic of interest. The chapter concludes with a brief comparison of these methods.

## Mental Model Theory

First defined by Bartlett (1932) and Craik (1943), a mental model is an internal, cognitive representation of the external world and how we understand it. Doyle and Ford (1999) describe the mental model as "… a relatively enduring and accessible, but limited, internal conceptual representation of an external system (*historical*, *existing*, or *projected*) [italics in original] whose structure *is analogous to* [italics in original] the perceived structure of that system" (p. 414). These internal representations are used by humans to understand and draw inferences about the world around them through mental simulations that are conducted on those internal representations (Craik 1943). Johnson-Laird builds on this notion and suggests that mental models are simplifications of real-world events, necessarily leaving out aspects of the external world which they simulate. Various theorists have aspired to provide a clear theory of mental models (Gentner 1983; Johnson-Laird 1983; Johnson-Laird and Bryne 1991; Norman 1983; Oakhill and Garnham 1996; Westbrook 2006). However, beyond consensus around the earlier principles, there is a wide variety of disparate concepts. Norman (1983) suggests mental models tend toward simplicity from limits on working memory. Johnson-Laird (1983) states that people make reasoning errors when mental models conflict because of this capacity limit, even though they can manipulate distinct mental models at the same time when explaining phenomena.

Diagrams and evaluations of stakeholder mental models allow decision-makers to see the influences that drive stakeholder behavior and their interactions with the systems that decision-makers are responsible for designing, implementing, and maintaining. The graphic representations can also facilitate empirical analyses of synergies and antagonisms between different perceptions and beliefs of different management options. Mental Modeling provides a way to summarize and compare views of management groups and stakeholder groups scientifically. Mental Modeling has been used in recent years, especially in the areas of human health and natural hazard mitigation (Chap. 4 this volume; Fischhoff et al. 2006). The influence diagrams produced from Mental Modeling give a simplified view of the domain of interest. Some methods for visually representing mental models provide good qualitative descriptions but make quantitative descriptions of risk difficult.

## Mental Model Diagrams

Authors use many different methods to represent mental models (Stevens and Gentner 1983). Mental models diagrams can help to show how people think about some topic. These diagrams can be compared across different stakeholders to identify similarities and differences (Carley 1997; Daniels et al. 1995). Mental models can be represented as propositional logic (Gentner 1983), as diagrams in the form of semantic webs (e.g., Novak and Gowin 1984), concept maps (Carley and Palmquist 1992; Valentine 1989), or influence diagrams (Clemen and Reilly 2001; Fischhoff et al. 2006). Visual representations of mental models can facilitate reasoning (Johnson-Laird 2006) and promote understanding of other stakeholders (Hoffmann 2005; Kolkman et al. 2005). As such, decision-makers may benefit from the reasoning support that mental model diagrams provide. Mental model diagrams can be elicited directly through targeted questioning (direct elicitation; Jones et al. 2011), or indirectly from respondents' verbalizations, decisions, and/or actions (indirect elicitation; Jones et al. 2011), or both.

## Mental Modeling History and Method

Mental Modeling first developed from a group of researchers at Carnegie Mellon University interested in using influence diagrams to model environmental risk decisions faced by the public as a first step in designing risk communications that better inform risk decisions (e.g., Bostrom et al. 1992; Darisi et al. 2005; Morgan et al. 2002). An influence diagram is a directional graph that represents the key variables of a system and their direction of influence. Mental Modeling has been used to evaluate risk communications for a variety of different hazards (e.g., Atman et al. 1994; Bostrom et al. 1994a, b; Gregory et al. 2003; Read et al. 1994). This method has also been used to inform strategic decision making (e.g., Scholz et al. 1999).

Mental Modeling is a six-step process that is described in detail in Chap. 2 and examples of the visualizations that Mental Modeling produces can be seen throughout this book. It produces an expert influence diagram with important concepts that influence a decision or system. It enables comparisons of systems of beliefs across stakeholder groups and can reveal important beliefs and knowledge absent from the expert model, identify conflicts and common ground across stakeholders, and facilitate risk communication design. Quantitative metrics can be used to compare the mental models of different stakeholders (Byram 1998; Vislosky and Fischbeck 2000). A particular strength of Mental Modeling is the ability to compare stakeholder mental models of a problem with all of their complexities, omissions, and misconceptions, to an explicit model of the state of the science as reified in the expert model. Comparisons with the expert decision model can illustrate opportunities for interventions that facilitate more effective decision making.

That said, Mental Modeling does have some limitations. It can be difficult to use in instances where the structure of stakeholder models differ (Bostrom et al. 1995). The method was built on a deficit model that compares complete expert knowledge to less complete layperson knowledge with a view to bringing layperson knowledge to expert competence. As structural differences between groups become more substantial, it becomes more difficult to compare the products of Mental Modeling to each other. In addition, content analysis (Neuendorf 2002) is typically carried out by hand with coding guidelines, although computational tools and algorithms are coming online for converting interview and questionnaire data into mental model diagrams (Chap. 14).

## Other Methods for Representing Mental Models

### *Concept Mapping*

Trochim (Trochim 1989a, b) developed concept mapping to combine mental models of different groups to inform decision planning in a variety of domains, in particular educational contexts (Galvin 1989; Pirnay-Dummer et al. 2008; Valentine 1989). Computer programs for concept mapping are currently available (Concept Mapping Resource Guide 2009; Ifenthaler 2010). The process in general has six steps, outlined in Table 3.1. The main product is a collection of 2D or 3D diagrams that represent semantic spaces where the distance between concepts is proportional to their similarity, with proximal concepts sharing greater similarity than distil concepts. Computer aids using this approach (e.g., Kearney and Kaplan 1997) have different metrics to assess concept relatedness, like the extent to which concepts are related (*closeness*) and the extent to which they differ (*contrast*; Pirnay-Dummer 2007).

Concept mapping provides a standardized approach to synthesizing stakeholder perspectives and can be performed by anyone with general experience in facilitation and some statistical expertise. This is also a limitation, however, since the method is optimized to present information from one stakeholder group and not for comparing across stakeholder groups. The quality of the concept map is more dependent on the constraints of facilitated sessions than other methods, especially focus statement

**Table 3.1** Concept mapping steps (from Trochim 1989a, b)

| Step | Description |
|---|---|
| Preparation | Select participants. Develop brainstorming focus statement and assessment criteria for brainstormed statements |
| Generation | Use focus statement to elicit statements from participants |
| Structuring | Conduct facilitator-guided statement sorting into as many self-defined categories as participants deem appropriate, with the constraint that (1) the total categories do not equal the number of statements and that (2) not all statements are grouped into one category. Collect participant ratings for each statement in terms of importance to the focus statement or expected outcome |
| Representation | Create a similarity matrix with category data. Create a series of maps with statistical techniques to represent similarity between statements. Show the relationship between statements with point maps, or the superordinate statement structure via cluster maps. Supplement with rating data to identify the relative importance of statements and clusters. |
| Interpretation | Present maps to participants to label categories and infer themes |
| Utilization | Use maps to identify key concerns and/or to evaluate a plan or program's success |

development. The majority of the process is participant driven and analyst facilitated, a benefit for fostering stakeholder engagement, but the black box nature of the statistical techniques employed may work against this benefit. The concept mapping method is good at quantifying the extent to which themes or categories are related, but does not suggest the nature or direction of influence between categories.

## *Semantic Web*

The semantic web method (Novak and Gowin 1984; Novak and Musonda 1991) was developed to assess improvements in children's understanding of science concepts over the course of instruction and has been used at all levels of education (Freeman and Jessup 2004). This method produces qualitative descriptions of relations between concepts from individual participants through guided elicitation at a higher resolution than other approaches, though this increased qualitative description can make numeric measurement difficult. Noun concepts are represented as nodes in a network and directional arrows labeled with functional relationships provide relational information between nodes.

Participants are asked, first at a high level, to describe some topic or scenario, and then to report on specific changes that would be effected if something were to change. The facilitator identifies segments from these transcripts and recordings that capture the participant's description of relationships between concepts and uses this information to create the influence diagram (Novak and Musonda 1991). This process can be done by individuals or small groups of participants, and can be drawn directly from participants when offline coding is intractable or direct elicitation is preferred for other reasons (Freeman and Jessup 2004). In both cases the diagrams are drawn so that nodes represent the key ideas for the topic of interest. Links between nodes have descriptions attached to them that capture how nodes are related. Some links are drawn as arrows to indicate a direction of influence, or simple lines can be used when nondirectional. Comparative statistics can be used to provide some quantitative measurement of semantic webs, though reported metrics may not generalize to new contexts (e.g., Novak and Musonda 1991).

Network diagrams created by semantic webs provide simple visual representations that can help communicate the relationships between concepts. Diagrams are intuitive and do not require training to interpret, and the functional terms that label links can help to create qualitative narrative descriptions of the diagram. Suggested scoring systems for measuring semantic webs are typically easy to calculate. However, semantic webs can be quite subjective and may be hard to use with larger groups. Since the technique was intended for measuring changes in mental models of participants as their expertise in an area increases, the approach may be less helpful for cross-participant comparisons.

### *System Dynamics Diagramming*

System dynamics diagramming provides a way to represent the behavior of complex systems from the problem owners' perspective. This approach considers a problem as a set of interrelated parts that can be deconstructed one from the other, where the interactions between parts can produce behavior that is different from individual components' behavior. The goal is to define the relationships between parts to describe the problem, and assume that the direction of influence between two parts is often bidirectional (Forrester 1961, 1993; Karkkainen and Hallikas 2006). Constraints from nonlinear dynamics and systems theory are used to guide the diagramming process and capture system behavior (Lane 2000). Many of the methods used for diagramming use complex math to quantify the relationships between components (Sterman 2001).

The system dynamics diagramming processes are loosely structured (Karkkainen and Hallikas 2006). First, the analyst defines the problem with participant assistance. The analyst and participants work together to identify causal factors and rules for transforming information to action. They work together to generate dynamic hypotheses that predict how parts of the system interact based on diagrams or other descriptions of participant mental models. Finally, the analyst builds the model with participant input, describing the parts of the system and any interactions between parts of the system mathematically. This model can then be used for simulation to understand the consequences of different management policies that are imposed. The models represented by system dynamics diagrams can be very complex and require computer simulations to effectively capture the complexity between parts of the system (e.g., Moxnes 2004). Systems dynamics research has been used in several instances to measure misunderstanding of relationships between different parts of mental models (Sterman 2008; Sterman and Sweeney 2007)

The system dynamics diagram process produces a model that can often simulate future scenarios based on input from participants. The representations generated in modeling exercises can help to capture complexity, especially complex relationships between concepts and groups of concepts. System dynamics models model quantitative differences in processes and outputs given specific inputs and allow participants to test different model inputs and management policies to identify optimal configurations. System dynamics diagramming can be quite useful for identifying and comparing relevant knowledge from different expert domains since the goal is to build simulations of systems, but this can also lead to a dizzying amount of complexity that is difficult to communicate.

## Conclusions

All of the earlier Mental Modeling approaches provide information about stakeholder beliefs and perceptions to varying degrees of conciseness and accessibility. The methods reviewed all provide visual representations of stakeholder beliefs, although how that information is elicited and visualized differs widely. This can create substantive differences in the final product's appearance and content. The best method for a specific task should be selected carefully depending on the goals of the analysis. Mental Modeling can complement many of these other techniques. One might use Mental Modeling to describe a risk management process, concept mapping to provide quantitative descriptions of certain aspects of that process, semantic web methods to increase descriptions of the influences between concepts, and system dynamics to simulate the system in action. That said, challenges for integration do exist. Each technique has processes for eliciting, collecting, and representing information, so information gained from participants for one modeling effort may be difficult to use for a complementary model in a different framework. Future research should develop complementary methods to use across these modeling frameworks and identify means to compare mental model diagrams of similar content with qualitatively different surface features.

## References

Atman, C. J., Bostrom, A., Fischhoff, B., & Morgan, M. G. (1994). Designing risk communications—completing and correcting mental models of hazardous processes. 1. *Risk Analysis, 14*(5), 779–788.

Bartlett, F. C. (1932). *A theory of remembering remembering: A study in experimental and social psychology* (pp. 197–214). London: Cambridge University Press.

Bostrom, A., Atman, C. J., Fischhoff, B., & Morgan, M. G. (1994a). Evaluating risk communications—completing and correcting mental models of hazardous processes. 2. *Risk Analysis, 14*(5), 789–798.

Bostrom, A., Fischbeck, P., Kucklick, J. H., & Hayward Walker, A. (1995). *A mental models approach to the preparation of summary reports on ecological issues related to dispersant use.* Washington, DC: Marine Spill Response Corporation.

Bostrom, A., Fischhoff, B., & Morgan, M. G. (1992). Characterizing mental models of hazardous processes: A methodology and an application to radon. *Journal of Social Issues, 4*, 85–100.

Bostrom, A., Morgan, M. G., Fischhoff, B., & Read, D. (1994b). What do people know about global climate change?: 1. Mental models. *Risk Analysis, 14*(6), 959–970.

Byram, S. (1998). *Breast cancer and mammogram screening: Mental models and quantitative assessments of beliefs*. Pittsburgh, PA: Carnegie Mellon University.

Carley, K. M. (1997). Extracting team mental models through textual analysis. *Journal of Organizational Behavior, 18*(Special Issue), 533–558.

Carley, K. M., & Palmquist, M. (1992). Extracting, representing, and analyzing mental models. *Social Forces, 70*(3), 601–636.

Clemen, R. T., & Reilly, T. (Eds.). (2001). *Making hard decisions with decision tools*. Pacific Grove, CA: Duxbury.

Concept Mapping Resource Guide. (2009). Retrieved July 9, 2008, from http://www.socialresearchmethods.net/mapping/mapping.htm#Papers

Craik, K. J. W. (1943). *The nature of explanation*. Cambridge, UK: University Press.

Daniels, K., de Chernatony, L., & Johnson, G. (1995). Validating a method for mapping managers' mental models of competetive industry structures. *Human Relations, 48*(9), 975–991.

Darisi, T., Thorne, S., & Iacobelli, C. (2005). Influences on decision-making for undergoing plastic surgery: A mental models and quantitative assessment. *Plastic and Reconstructive Surgery, 116*(3), 907–916.

Downs, J. S., Bruine de Bruin, W., & Fischhoff, B. (2008). Parents' vaccination comprehension and decisions. *Vaccine, 26*, 1595–1607.

Downs, J. S., & Fischhoff, B. (2011). Qualitative information. In B. Fischhoff, N. T. Brewer, & J. S. Downs (Eds.), *Communicating risks and benefits: An evidence-based user's guide*. Washington, DC: Food and Drug Administration.

Doyle, J. K., & Ford, D. N. (1999). Mental models concepts revisited: Some clarifications and a reply to Lane. *System Dynamics Review, 15*(4), 411–415.

Fischhoff, B. (1995). Risk perception and communication unplugged: Twenty years of process. *Risk Analysis, 15*(2), 137–145.

Fischhoff, B., de Bruin, W. B., Guvenc, U., Caruso, D., & Brilliant, L. (2006). Analyzing disaster risks and plans: An avian flu example. *Journal of Risk and Uncertainty, 33*(1-2), 131–149.

Forrester, J. W. (1961). *Industrial dynamics*. New York: The MIT Press and John Wiley & Sons, Inc.

Forrester, J. W. (1993). System dynamics and the lessons of 35 years. Unpublished manuscript.

Freeman, L. A., & Jessup, L. M. (2004). The power and benefits of concept mapping: Measuring use, usefulness, ease of use, and satisfaction. *International Journal of Science Education, 26*(2), 151–169.

Galvin, P. F. (1989). Concept mapping for planning and evaluation of a Big Brother/Big Sister program. *Evaluation and Program Planning, 12*, 53–57.

Gentner, D. (1983). A theoretical framework for analogy. *Cognitive Science, 7*, 155–170.

Gregory, R., Fischhoff, B., Thorne, S., & Butte, G. (2003). A multi-channel stakeholder consultation process for transmission deregulation. *Energy Policy, 31*(12), 1291–1299.

Hoffmann, M. H. G. (2005). Logical argument mapping: A method for overcoming cognitive problems of conflict management. *International Journal of Conflict Management, 16*(4), 304–334.

Ifenthaler, D. (2010). Relational, structural, and semantic analysis of graphical representations and concept maps. *Educational Technology Research and Development, 58*(1), 81–97.

Johnson-Laird, P. N. (1983). *Mental models*. Cambridge: Harvard University Press.

Johnson-Laird, P. N. (2006). Models and heterogeneous reasoning. *Journal of Experimental & Theoretical Artificial Intelligence, 18*(2), 121–148.
Johnson-Laird, P. N., & Bryne, R. M. J. (1991). *Deduction*. Hillsdale, NJ: Lawrence Earlbaum Associates.
Jones, N. A., Ross, H., Lynam, T., Perez, P., & Leitch, A. (2011). Mental models: An interdisciplinary synthesis of theory and methods. *Ecology and Society, 16*(1), 46.
Karkkainen, H., & Hallikas, J. (2006). Decision making in inter-organisational relationships: Implications from systems thinking. *International Journal of Technology Management, 33*(2-3), 144–159.
Kearney, A. R., & Kaplan, S. (1997). Toward a methodology for the measurement of knowledge structures of ordinary people—the conceptual content cognitive map (3CM). *Environment and Behavior, 29*(5), 579–617.
Kiker, G. A., Bridges, T. S., Varghese, A., Seager, T. P., & Linkov, I. (2005). Application of multi-criteria decision analysis in environmental decision making. *Integrated Environmental Assessment and Management, 1*(2), 95–108.
Kolkman, M. J., Kok, M., & van der Veen, A. (2005). Mental model mapping as a new tool to analyse the use of information in decision-making in integrated water management. *Physics and Chemistry of the Earth, 30*(4-5), 317–332.
Lane, D. C. (2000). Should system dynamics be described as a "hard" or "deterministic" systems approach? *Systems Research and Behavioral Science, 17*(1), 3–22.
Linkov, I., Cormier, S., Gold, J., Satterstrom, F. K., & Bridges, T. (2012). Using our brains to develop better policy. *Risk Analysis, 32*(3), 374–380.
Morgan, M. G., Fischhoff, B., Bostrom, A., & Atman, C. (2002). *Risk communication: A mental models approach*. New York: Cambridge University Press.
Moxnes, E. (2004). Misperceptions of basic dynamics: The case of renewable resource management. *System Dynamics Review, 20*(2), 139–162.
Neuendorf, K. A. (2002). *The content analysis guidebook*. Thousand Oaks, CA: Sage Publications.
Norman, D. A. (1983). Some observations on mental models. In D. Gentner & A. L. Stevens (Eds.), *Mental models* (pp. 7–14). Hillsdale, NJ: Lawrence Earlbaum Associates.
Novak, J. D., & Gowin, D. B. (1984). *Learning how to learn*. New York: Cambridge University Press.
Novak, J. D., & Musonda, D. (1991). A twelve-year longitudinal study of science concept learning. *American Educational Review Journal, 28*(1), 117–153.
Oakhill, J., & Garnham, A. (Eds.). (1996). *Mental models in cognitive science*. Mahwah, NJ: Lawrence Erlbaum Associates.
Pirnay-Dummer, P. (2007, April). *Model inspection trace of concepts and relations: A heuristic approach to language-oriented model assessment*. Paper presented at the annual meeting of American Educational Research Association, Chicago, IL.
Pirnay-Dummer, P., Ifenthaler, D., Johnson, T., & Al-Diban, S. (2008, March). *Reading with the guide of automated graphical representations: How model based text visualizations facilitate learning in reading comprehension tasks*. Paper presented at the annual meeting of the American Educational Research Association, New York, NY.
Read, D., Bostrom, A., Morgan, M. G., Fischhoff, B., & Smuts, T. (1994). What do people know about global climate change? 2. Survey studies of educated laypeople. *Risk Analysis, 14*(6), 971–982.
Scholz, D. K., Kucklick, J. H., Pond, R., Walker, A. H., Bostrom, A., & Fischbeck, P. (1999). *Fate of spilled oil in marine waters: Where does it go, what does it do, and how do dispersants affect it?* Cape Charles, VA: Scientific and Environmental Associates, Inc.
Sterman, J. D. (2001). System dynamics modeling: Tools for learning in a complex world. *California Management Review, 43*(4), 8–25.
Sterman, J. D. (2008). Economics—risk communication on climate: Mental models and mass balance. *Science, 322*(5901), 532–533.

Sterman, J. D., & Sweeney, L. B. (2007). Understanding public complacency about climate change: Adults' mental models of climate change violate conservation of matter. *Climatic Change, 80*(3-4), 213–238.

Stevens, A. L., & Gentner, D. (1983). Introduction. In D. Gentner & A. L. Stevens (Eds.), *Mental models* (pp. 1–6). Hillsdale, NJ: Lawrence Erlbaum Associates.

Trochim, W. M. K. (1989a). Concept mapping: Soft science or hard art? *Evaluation and Program Planning, 12*, 87–110.

Trochim, W. M. K. (1989b). An introduction to concept mapping for planning and evaluation. *Evaluation and Program Planning, 12*, 1–16.

Valentine, K. (1989). Contributions to the theory of care. *Evaluation and Program Planning, 12*, 17–23.

Vislosky, D. M., & Fischbeck, P. S. (2000). A mental model approach applied to R&D decision-making. *International Journal of Technology Management, 19*(3-5), 453–471.

Westbrook, L. (2006). Mental models: A theoretical overview and preliminary study. *Journal of Information Science, 32*(6), 563–579.

# Part II
# Applications at U.S. Army Corps of Engineers (USACE)

# Chapter 4
# Flood Risk Management

**Matthew D. Wood, Igor Linkov, Daniel Kovacs, and Gordon Butte**

## Introduction

The disastrous hurricane season in 2005 exposed several weaknesses in the storm and flood management plans along the U.S. coastline of the Gulf of Mexico. Specifically, Hurricane Katrina in late August 2005 and Hurricane Rita in mid-September 2005 placed immense strain on the engineering design and coastal management of many Gulf States, causing many to blame government response and planning for extreme storm and flood situations (such as levee construction and wetlands management; Cigler 2007; Johnson 2005). While most focused on these

---

This chapter is adapted from work by Wood et al. (2012). Flood risk management: U.S. Army Corps of Engineers and layperson perceptions. Risk Analysis, 32(8), 1349–1368. Special thanks to Dr. Todd Bridges, Senior Research Scientist, U.S. Army Corps of Engineers, Engineer Research and Development Center, for his contributions to this chapter.

M.D. Wood, Ph.D. (✉)
U.S. Army Corps of Engineers, Engineer Research and Development Center (ERDC) and Carnegie Mellon University, 696 Virginia Road, Concord, MA 01742, USA
e-mail: Matthew.d.wood@usace.army.mil

I. Linkov, Ph.D.
U.S. Army Corps of Engineers, Engineer Research and Development Center (ERDC) and Carnegie Mellon University, 696 Virginia Road, Concord, MA 01742, USA
e-mail: Igor.Linkov@usace.army.mil

D. Kovacs, Ph.D.
Decision Partners, 1458 Jersey Street, Lake Milton, OH 44429, USA
e-mail: dkovacs@decisionparters.com

G. Butte
Decision Partners LLC, Suite 200, 313 East Carson Street, Pittsburgh, PA 15217, USA
e-mail: gbutte@decisionpartners.com



M.D. Wood et al., *Mental Modeling Approach*, Risk, Systems and Decisions,
DOI 10.1007/978-1-4939-6616-5_4

factors as the primary driver behind the extensive damage to the Gulf Coast, recent inquiries have also discussed the role of other factors in extreme storm and flood planning, such as the importance of individual psychological factors of citizens to disaster management planning (Gheytanchi et al. 2007). Additional recommendations were issued by a discussion panel of The Institution of Civil Engineers prior to the dual-hurricane disaster. In their response, the Institution recommended that government agencies develop waterway expansion plans to accommodate rainfall and account more for human and societal factors during flood and extreme weather situations (Fleming 2002a, b).

Since 2005, the U.S. Army Corps of Engineers (USACE) and similar U.S. government agencies have promoted interagency cooperation and inclusion of local stakeholders in their efforts to improve and restore coastal areas of Louisiana and Mississippi (National Research Council 2008; USACE 2008; Working Group for Post-Hurricane Planning for the Louisiana Coast 2006). Specifically, USACE and others seek to bolster stakeholder participation in these efforts by identifying stakeholder needs and goals for the region and incorporating these tasks into restoration efforts. In this effort, USACE seeks to shift from its focus on engineering for coastal protection to adopting coordination and cooperation strategies with authorities in state and local government. In this framework, state and local government stakeholders can help to improve information and coordination to coastal planning in ways that would not have otherwise been considered, thereby benefitting the local community (Hecker et al. 2008; Rabbon et al. 2008). Such state and local involvement can also help USACE in its duty to monitor and maintain 383 lakes and reservoirs, 8500 miles of dikes and levees, and over 240 miles of shoreline protection projects that require frequent upkeep (U.S. Army Engineer Institute for Water Resources 2009).

One recurring complication for government agencies is the expectation by many private citizens that these agencies will shoulder the task of flood risk management, even regarding the issues of loss prevention and personal insurance (Gheytanchi et al. 2007; Vari et al. 2003, Lave and Lave 1991). This mind-set is further perpetuated by the substantial infrastructure and architecture networks developed by USACE and others to counter a flood event alongside the National Flood Insurance Program (NFIP) requirement, where many assume that personal responsibility and risk management are not required or necessary for the private citizen (Hecker et al. 2008; Rabbon et al. 2008). This is driven by multiple factors, including the general citizen's difficulty in recalling the specifics of past flood events and a strong underestimation of the impact and likelihood that such a flood event could arise in the near future. While USACE's National Flood Risk Management Program seeks to address some of these issues, further involvement by stakeholders and local citizens in flood risk management planning is needed to overcome remaining preconceptions by citizens that their active participation is not necessary in flood risk situations (Hecker et al. 2008; Rabbon et al. 2008).

Along with its efforts to develop stakeholder involvement and participation in flood risk management efforts, USACE and others are also working to bolster interagency coordination and cooperation in flood risk planning and risk management. Such interagency synchronization requires significant effort to organize and gain the participation of agencies such as FEMA, DHS, and a collection of various other federal, state, and local authorities for a multitude of flood risk projects and policies. Further complicating this situation include the difficulties of integrating the opinions, knowledge, and value assessments of various public and private stakeholders in a more formalized manner, where ad hoc measures to aggregate such information can easily lead to clutter and disorganization of policy measures and priorities. One such method to overcome this problem is Mental Modeling, which can facilitate improved understanding of deeply held risk and value beliefs in specific situations. For flood risk management, the use of Mental Modeling may help further stakeholder involvement in planning and execution, particularly where involving the everyday private citizen is concerned. Though policy experts, government officials, and private citizens have been demonstrated as having complex sets of values and differing opinions that are difficult to collectively address, we seek to demonstrate that Mental Modeling serves as a method to integrate a variety of opinions and inputs to advance flood risk management and policy goals (Lave and Lave 1991; Kolkman et al. 2007; Wagner 2007).

This chapter discusses perceptions of various USACE expert and lay stakeholders on flood risk management. Particularly, this chapter focuses on how different experts within USACE comprehend and understand flood risk management, and subsequently identifies potential areas for improvement in such areas from a policy perspective. One resulting contribution by this study includes an assessment of the differences in flood risk management perceptions and opinions by USACE engineers and policy planners, respectively, where we utilize information gathered from Mental Modeling interviews to serve as data for quantitative analysis. These two groups of USACE personnel are tasked with very different responsibilities in the flood risk management process, and a comparison of their expert judgments can help to illuminate areas for refining or improving the flood risk management process. Ultimately, this study's results will assist USACE's future flood risk management planning process and will help indicate to other agencies the types of activities and tools needed to improve flood risk policy by incorporating a variety of stakeholder beliefs and viewpoints.

To accomplish these goals, we assess the opinions and beliefs of two stakeholder groups within USACE on flood risk management. Additionally, we also make use of existing published research on the opinions of private citizens toward flood risk, including those containing mental models of flood risks and flood risk management. In this way, we demonstrate how we can integrate qualitative and quantitative information from existing scholarship while gathering new qualitative information from subject matter expert interviews. This allows for an improved comparison of flood risk management views across different stakeholders, where Mental Modeling serves as a formal method to assess subjective opinions and integrate such information in a transparent and quantitative manner. This chapter makes use of a literature review to understand how layperson perceptions should be considered when improving the USACE FRM process and mental models interviews with USACE FRM experts to understand expert mental models of flood risk and how current processes can be improved.

## Literature Review of Layperson Stakeholder Perceptions

In our first step to compare lay versus expert flood risk management opinions and perceptions, we first initiated a literature review in order to better understand the mental models of private citizens and non-USACE subject experts. Conducted May–July 2008, this literature review was motivated by the need to understand how USACE flood risk management policies and operations may take into account the opinions and preferences of other stakeholders in the flood risk management process. The review began via a keyword search in the ISI Web of Knowledge for the terms "mental model," "flood manage," "flood respon*," and "flood recov*." Additional keyword searches were conducted on ISI and PSYCInfo with additional relevant terms discovered in the initial literature search, including "decision support systems," "concept mapping," "strategic environmental assessment," "Logical Argument Mapping," "risk-ranking," "shared mental model," and "team mental model." A more limited search was conducted afterward using the keywords "schema*" and "folk model*." Unfortunately, these searches generated a large number of unrelated and irrelevant results. Steps were taken to refocus results on flood risk management, including those papers which cited or were aligned with definitions of mental models presented in Morgan et al. (2002) or Doyle and Ford (1999). The subsequent works offered a clear definition for mental models of complex real-world phenomena.

Queries identified by ISI Web of Knowledge, PSYCInfo, and Social Science Citation Index "Expanded Searches" were also supplemented with reviews of query reference sections to identify sentinel works, and a Google Scholar search of key authors to locate books or other published works on the topic not available via peer-reviewed journal sources. Ultimately, our literature search produced over 600 published documents. Of these documents, select articles that provided information on stakeholder perceptions of the flood risk management process were retained for future use in this case study. Additionally, the document had to match one or more components of the Expert Model (Fig. 4.1). Characteristics that excluded an article from further use included those which detailed existing risk management decision support systems or had an excessively extensive focus on the technical elements of hydrological, meteorological, or geological processes that implicated flooding. Overall, it was necessary to remove these types of articles from further discussion, as they did little to further the understanding of lay stakeholder opinions and beliefs in flood risk management situations.

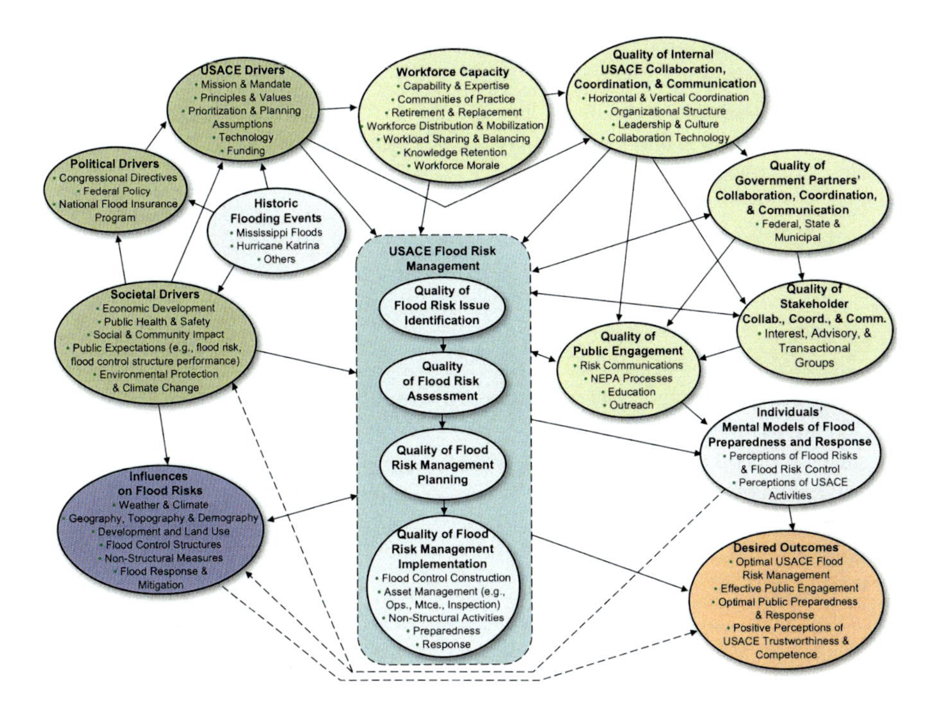


**Fig. 4.1** Simple expert model of influences on USACE food risk management

## Literature Review Results

Though this review was not necessarily exhaustive, it does provide information regarding the policy problems and concerns facing organizations that are required to perceive and manage flood risk events. Specifically, the literature review results provide lay stakeholder perspective information related to *Societal Drivers, Influences on Flood Risks, Quality of FRM Planning, and Quality of FRM Implementation* (see Expert Modeling later).

For the first element, societal drivers are those factors which are created by society that have some influence on the flood risk management process. As mentioned earlier, the drive and willingness to be preemptively engaged in a flood event and adequately prepare for such future scenarios are chief societal drivers for flood risk management. The ability to acquire and distribute accurate information is the central component toward bolstering positive societal drivers, such as with the provision of accurate weather forecasting with enough time for those affected by floods to prepare for them (Regnier and Harr 2006). Another important societal driver includes a coordination and clear line of communication between private citizens and experts, where past publications note that potential disconnects between these groups on the subject of flood preparation can slow or hinder the process of reducing public exposure to flood risk (Pielke 1999). This driver is a particularly challenging one, as lay stakeholders generally know little regarding how to decrease their own exposure to flood risk and do not immediately know how to acquire such information. This is a case of ignorance of what to do rather than a rejection of plans developed by experts that results in laypersons ignoring the advice of flood control experts (Lave and Lave 1991).

In this vein, the provision of information of the ground, along with an accurate and regular forecasting of future events, is an important activity to be undertaken by experts and relayed in an effective way to private citizens (Rasid and Haider 2002). Buckland and Rahman describe the benefits of strong communal social capital in flood risk situations, where they note that those societies with higher levels of physical, human, and social capital were better prepared for hazardous flood events (Buckland and Rahman 1999). However, it should be noted that higher levels of social capital made the flood preparation process somewhat more difficult and contributed to delays in the decision process. Overall, though, elevated social and human capital does have a tendency to bring communities together in an organized manner generally not seen in those communities with low social, human, and physical capital (Clark et al. 2002).

With respect to influences on flood risks, many factors are described as having direct and potentially drastic influences on the outcomes of flood risk scenarios. Of these, perhaps the most salient and consequential factors that arose in the literature reviewed included *Weather & Climate, Geography, Tomography & Demography,* and *Development and Land Use.* For example, Lave and Lave (1991) discussed the notion that those in flood-prone areas are ill informed of their actual flood risk or how flood events initiate, and have little enthusiasm or motivation to acquire such information. In addition, these inhabitants often greatly overestimate the ability of experts to predict the arrival of such flood events in a timely manner, and that any recent floods with little warning and significant damage are anomalous. Higher educated citizens who self-identified as being scientifically proficient tend to know much more about the flooding process than their ill-informed peers and are better prepared for such hazards (Pielke 1999).

An additional common belief among lay citizens includes skepticism of the government's ability to manage water resources in an effective manner. Where waterways are generally organized and managed on a local level and generally not as a piece of a larger waterway management strategy, widespread coordination of waterway resources in flood events can be a complex challenge. Some respondents to Lave and Lave (1991) blamed flooding events in their area on poor coordination and regional waterway management operations, where government efforts and focus was perceived to center on preserving urban areas from flood risk instead of some suburban or rural communities. In this vein, many perceive their local water management efforts of managing flood risk as a trade-off between benefits for certain groups (the privileged) and hazards for others (the less privileged; USACE 2008).

Overall, limited knowledge or understanding associated with flood events, likelihood, or protective measures are likely to increase the potential costs in any resulting flood damages. This is only heightened by the fact that many who live in flood plains overestimate the capabilities of physical flood barriers such as levees and assume that these structures will prevent damages (Pielke 1999). While levees and other physical structures may reduce the near-term likelihood of certain flood events, they cannot completely eliminate the potential for such an event to occur—particularly with respect to more serious floods. Instead, these structures have the potential to negatively influence the local wetlands or water table, which may increase the potential for a more damaging flood in years to come (Fischetti 2001). As such, the line of thinking for many without an accurate understanding of flood risk management is "Am I Safe?" when a more accurate view would be "How Safe Am I" from a flood event? (Johnson and Slovic 1998).

Another factor which has a dramatic effect on flood outcomes includes flood risk management implementation, where the ability of expert groups to communicate with each other drives the local, state, and federal governments' ability to take appropriate action to mitigate flood risk. Such communication across groups is a tricky and complex process and is rife with differing technical terminology (reference terms or jargon used by different groups to reference similar concepts), political drivers, uncertain timelines, and legal responsibilities. With respect to technical terminology, one example of such differences includes an inability of professional emergency managers to accurately assess flood uncertainties provided by meteorologists without direct guidance and assistance (Faulkner et al. 2007). Lave and Lave also point out the difficulties associated with short memories and an inability of experts to communicate long-term flood risk potential to the public, where private citizens are likely to purchase flood insurance after a damaging event yet cancel such insurance after a few years of no harmful events (Lave and Lave 1991). Collectively, these flood risk management implementation concerns require clear and transparent coordination between different expert groups alongside an engaged public.

## Expert Modeling

The expert modeling approach deployed in this case study is used to assess beliefs about flood risk management and has been applied in the past to a variety of risk management interests and domains (Morgan et al. 2002; Downs et al. 2008; Wood et al. 2009; Thorne et al. 2005). Historical Mental Modeling efforts with some degree of success with environmental policy and management concerns include projects contrasting expert and layperson knowledge regarding climate change (Lowe and Lorenzoni 2007; Read et al. 1994; Bostrom et al. 1994), environmental manager views on climate change and biodiversity (Hill and Thompson 2006), public perceptions of carbon sequestration techniques (Palmgren et al. 2004), and layperson perspectives of flash floods and landslides (Hill and Thompson 2006).

Within the Mental Modeling process, an important step in visualizing the process in question is to develop an *expert model*, which is a graphical representation of a mental model influence diagram (Hill and Thompson 2006). Such a diagram displays variables relevant to the decision process at hand, and may offer some assessment of the direction and strength of association of one variable's influence on another throughout the system. For this case study, an expert model was developed with a small number of USACE experts and was further developed into a more substantial expert model via a workshop with 11 USACE experts in a September 2008 workshop and subsequent follow-up interviews (Linkov et al. 2009). This EM served as the basis for developing a semistructured mental model interview protocol, with the ultimate objective of acquiring further information into the specific drivers and variables associated with USACE flood risk management. Additionally, this mental model interview protocol sought to elucidate any differences in perspective between planners and engineers within the USACE community.

## Expert Modeling Results

The *Expert Model of Influences on USACE Flood Risk Management* (Fig. 4.1) is illustrated as an influence diagram (Fleming 2002a, b) or a directed graph where different nodes (variables) are connected by arrows (a marker of influence between two variables) in a larger system. In this case, arrow directionality denotes the direction of influence, or otherwise marks which variable has an impact on the other. Following, we discuss the Expert Model in further detail.

The Model begins in the top-left corner with *Political, Social,* and *USACE Internal Drivers*, those factors that establish the fundamental conditions that influence USACE activities. Political Drivers include those elements that originate from government organizations that influence or have some form of regulatory control of USACE flood risk management activities, and include examples such as *Congressional Directives, Federal Policy,* and *NFIP* (which directly mandates flood insurance in flood-prone areas). Social Drivers include those factors which originate from society at large, and include *Economic Development Priorities, Public Health & Safety, Social & Community Impact* (the need to protect social networks and community organizations), *Public Expectations* (regarding flood risk and control), and *Environmental Protection & Climate Change.* USACE Drivers are internal factors within USACE that have some influence on their flood risk management capabilities and duties, including *Funding, Mission & Mandate, Principles & Values Prioritization of Flood Risk Management,* and *Planning Assumptions.* All of these drivers are directly influenced by historical flood events, particularly extreme cases as with the aftermath of Hurricane Katrina and the flooding along the Mississippi River Valley.

The notes in the middle of the Model relate to *USACE Flood Risk Management* activities, which are a visualization of the actual activities undertaken by USACE personnel. Specific activities which have an impact on *Quality of Flood Risk Management Implementation* include *Flood Control Construction* (building dams and levees), *Asset Management, Non-Structural Activities, Flood Preparedness,* and *Flood Response.* Collectively, these activities influence the lower left corner nodes entitled *Influences on Flood Risks*, which include naturally occurring variables such as weather and geography alongside direct human influences as with flood control structures, policies, and decision making. This collection of nodes includes influences of *Weather & Climate, Geography, Topography & Demography, Development and Land Use, Flood Control Structures, Non-Structural Measures,* and *Flood Response and Mitigation* (emergency repair of damaged structures and levees, flood drainage, and evacuation).

The nodes in the top right section of the model include those factors which influence USACE activities related to *Workforce Capacity* and the Corps' *Collaboration, Coordination, and Communications* alongside *Quality of USACE Public Engagement.*

Ultimately, the Model endpoint centers on the *Desired Outcomes* node in the bottom right-hand corner. This node collectively indicates the ability of USACE flood risk management activities to achieve their designated goals and fulfill necessary objectives. These goals and objectives include:

1. Optimal USACE flood risk management.
2. Effective public engagement.
3. Optimal public preparedness and response.
4. Positive perceptions of USACE trustworthiness and competence.

## Mental Models Interviews and Comparative Analysis

The last element of research for this case study includes the mental model interviews and subsequent comparative analysis, which is driven and influenced by the literature review and expert models discussed earlier. The interview protocol was constructed using influences from the expert model in order to fashion a semistructured interview protocol. The goal of these interviews was to gain a better understanding of the specific drivers on the quality of USACE flood risk management alongside an improved perspective on the differences in flood risk management belief and opinion between USACE planners and engineers, respectively. Additionally, we were particularly interested in determining the differences between these two groups regarding the drivers they emphasize or place particular value upon.

To accomplish these goals, we interviewed USACE engineers ($n=10$), planners ($n=8$), and program managers ($n=4$). These interviews were conducted over the telephone. Due to the similarities in mission and job function, we considered the program managers in this case to fall under the "planner" category. Overall, the interview subjects represented a convenience sample of senior USACE personnel holding positions in various units of the Corps.

The interview process and protocol were generated by using previous work on USACE personnel beliefs and opinions in the flood risk management process as an example. Changes to this older protocol were minor. Due to the general consistency of earlier engineer participants from an October 2008 interview survey, data from five engineers is included in our analysis of this case study. We were particularly interested in information related to:

1. The most important drivers of USACE flood risk management.
2. The scope and quality of USACE flood risk management activities.
3. Drivers of the quality of collaboration, coordination, and communication.
    (a) Internally within USACE.
    (b) Externally with other government, NGO, and lay stakeholders.

With these lines of inquiry in mind, interviewers asked questions to assess key features, opinions, and perceptions of the factors listed earlier by USACE personnel as related to flood risk management (Yoe 1993). To capture a full response, questions were worded broadly for an individual topic initially, and then narrowed and focused on a series of specific elements associated with the topic. This interview style and structure is consistent with past interview efforts developed by expert model diagrams (Morgan et al. 2002; Thorne et al. 2005). Interviews lasted approximately 1 h on average (mean 66 min; range 45–82 min) and were collected from October 2008 through June 2010. All discussions were recorded for future reference and subsequently transcribed.

These interviews were then analyzed via content analysis, whereby questions and their responses were reviewed individually and assessed for common themes across respondents. For consistency, each response was put through a two-pass process. In the first pass, each response by each individual respondent was read and associated with particular elements of the Expert Model. This facilitated comparison of expert model topics across all interviews and all interview questions. The second pass included a more thorough analysis, where each response was analyzed for more fundamental themes that could be identified within topic areas associated with the expert model. We were particularly interested in relative frequencies across participants that discussed a particular theme. Due to the sample size, differences between cohorts are reported here only if they are greater than 20 % (coding is rounded to the nearest 5 % by topic frequency).

## Interview Results

When asked about drivers to flood risk management activities, interview respondents answered with a variety of factors consistent with the Expert Model. The respondents discussed, in particular, the drivers related to public safety and reducing risk to the public (55 %), which is a primary element of USACE's core mission. Also frequently mentioned were Congressional mandate, funding, and politics (45 %) alongside historical events and reactive instead of proactive protection (45 %). Several suggested that Congressional mandate and politics were related to funding, where project funding is derived from Congressional approval. In this vein, respondents described an environment where it is sometimes difficult to get funding for projects (especially large projects) unless a catastrophe has occurred, even when USACE personnel feel the potential for such a catastrophe is an impending threat.

A sizeable fraction of respondents mentioned public interests in their responses, with 30 % speaking directly to local interest, support, sponsorship, and cost sharing. Economics, National Economic Development (NED), and economic efficiency (25 %)

also served as a common theme of discussion of drivers that influenced the USACE flood risk management process. Respondents also noted the tendency for localities to absorb a disproportionate amount of flood risk management cost, where the cost/benefit ratio for project selection and adoption is determined on a national scale.

Overall, engineers tended to discuss the five themes related to flood risk management drivers more so than the planners did in their responses. Engineers, in particular, focused on Congressional mandates and political drivers (60 % of engineers versus 35 % of planners). Several engineers stated that Congressional mandates drive current USACE objectives, particularly for big projects with large budget requests and resource needs. Collectively, engineers and planners cited politics and congressional mandates (40 %), economic mandates (30 %), and response to weather and significant effects (25 %) as the most significant drivers behind flood risk management activity. A smaller number of respondents mentioned local interests, environmental protection/impact, and technological drivers, although to a lesser extent. Planners were significantly more likely to stress politics and congressional mandates, while engineers centered on economic development and responses to weather and significant events. Due to the organizational roles and differences between these two groups, this disparity is sensible, as planners are more likely to seek project financing while engineers directly conduct site visits and work within areas recently or currently affected by natural events and weather hazards.

Regarding areas of improvement, engineers especially noted the gap between public expectation and reality (90 % of engineers versus 40 % of planners) associated with flood risk management. This was particularly true for concerns associated with the degree of protection that a structure may offer from flood events over the longer time horizon. Planners and engineers collectively stressed a need for improved public education on flood risk, where they stated that simple misunderstandings of statistical terms and probabilistic events such as "the hundred-year flood" make the public less attuned to actual flood hazards. Engineers also suggested that existing project evaluation criteria were problematic (70 % of engineers). Some found NED to be an unfair scoring metric, as it calculates costs for both federal and local groups, but only focuses on national benefits. Others also focused on the uncertainty regarding how to use the variety of alternative metrics for project planning and assessment. In contrast, planners tended to focus on the need to improve collaboration internally within USACE and externally with partner organizations (40 %). Planners also emphasized the need to reduce long project timelines (35 %). Several planners stated that improved collaboration may result in an improvement in project timelines, although they recognized that this would be a significant regulatory and cultural barrier to overcome.

## Discussion

The collective literature review, Expert Model, and mental model interviews suggests lay stakeholders remain unaware of flood risk causes and measures that may be taken to mitigate or manage flood risk. This is despite recent flooding catastrophes. Several authors noted in the literature review provide suggestions to close this gap, many of which could greatly assist USACE and its partner government agencies in their flood risk management preparation process.

The literature review and Expert Model collectively stress the influence of stakeholder perceptions have on flood risk management desired outcomes as associated with USACE's flood risk management goals. The Expert Model also suggests that individual mental models can be brought into alignment with those of experts in two general ways:

1. Directly from improvements to USACE's flood risk management process.
2. Indirectly through improving the quality of public engagement.

In the direct option, USACE can promote initiatives to resolve differences between public understanding, perceptions of USACE's flood risk management process, and realistic expectations of both the FRM process and flood risks. Likewise, the indirect option seeks to improve public involvement in the flood risk management process, particularly by improving flood risk education and participation.

Interview respondents noted USACE's ongoing efforts to work collaboratively with external stakeholder groups, but also described shortcomings of this process due to money and manpower limitations. They also described efforts to improve internal communication and collaboration across USACE agencies. However, efforts to work with external partners also foster internal technical and operational problems. Even so, several respondents noted that internal communication and collaboration were in a good state, and any issues were due to individual working groups or individual personnel rather than a system-wide shortcoming. The primary noted limitation is a lack of resources to pursue these ends.

As USACE flood risk management is a continuously evolving topic with a variety of drivers, future work should continue to incorporate research that reviews various expert and lay stakeholder perceptions to improve collaboration and participation in the flood risk management process. The use of decision analysis and Mental Modeling methods may also help in utilizing various sources of information in literature, expert modeling, and additional interviews with lay and subject expert stakeholders. Additionally, more specific information regarding to how the various drivers in the flood risk management process could help to suggest future action and avenues for improvement by USACE in its efforts to involve the public in its flood-related projects.

## References

Bostrom, A. M., Morgan, G., Fischhoff, B., & Read, D. (1994). What do people know about global climate change? 1. Mental models. *Risk Analysis, 14*(6), 959–979.

Buckland, J., & Rahman, M. (1999). Community-based disaster management during the 1997 Red River Flood in Canada. *Disasters, 23*(2), 174–191.

Clark, D. E., Novotny, V., Griffin, R., Booth, D., Bartosova, A., Daun, M. C., et al. (2002). Willingness to pay for flood and ecological risk reduction in an urban watershed. *Water Science and Technology, 45*(9), 235–242.

Cigler, B. A. (2007). The "big questions" of Katrina and the 2005 great flood of New Orleans. *Public Administration Review, 67*, 64–76.

Thorne, S. L., Darisi, T., & Iacobelli, C. (2005). Influences on decision-making regarding having plastic surgery: A mental models and quantitative assessment: P32. *Plastic and Reconstructive Surgery, 116*(3), 158–159.

Downs, J. S., Bruine de Bruin, W., & Fischhoff, B. (2008). Parents' vaccination comprehension and decisions. *Vaccine, 26*, 1595–1607.

Doyle, J. K., & Ford, D. N. (1999). Mental models concepts revisited: Some clarifications and a reply to Lane. *System Dynamics Review, 15*(4), 411–415.

Faulkner, H., Parker, D., Green, C., & Beven, K. (2007). Developing a translational discourse to communicate uncertainty in flood risk between science and the practitioner. *Ambio, 36*(8), 692–703.

Fischetti, M. (2001). Drowning New Orleans. *Scientific American, 285*(4), 76–85.

Fleming, G. (2002a). How can we learn to live with rivers? The findings of the Institution of Civil Engineers Presidential Commission on flood-risk management. *Philosophical Transactions of the Royal Society of London A: Mathematical, Physical and Engineering Sciences, 360*(1796), 1527–1530.

Fleming, G. (2002b). Learning to live with rivers—the ICE's report to government. *Proceedings of the ICE-Civil Engineering, 150*(5), 15–21.

Gheytanchi, A., Joseph, L., Gierlach, E., Kimpara, S., Housley, J., Franco, Z. E., & Beutler, L. E. (2007). The dirty dozen: twelve failures of the Hurricane Katrina response and how psychology can help. *American Psychologist, 62*(2), 118.

Hecker, E. J., Zepp, L. J., & Olsen, J. R. (2008). *Improving public safety in the United States—from federal protection to shared flood risk reduction. Presented at the European Conference on Flood Risk Management*. Oxford, UK: Keble College.

Hill, S. D., & Thompson, D. (2006). Understanding managers' views of global environmental risk. *Environmental Management, 37*(6), 773–787.

Johnson, B. L. (2005). Hurricane Katrina and that vexing 'what if?' question. *Human and Ecological Risk Assessment, 11*(6), 1081–1082.

Johnson, B. B., & Slovic, P. (1998). Lay views on uncertainty in environmental health risk assessment. *Journal of Risk Research, 1*(4), 261–279.

Kolkman, M. J., van der Veen, A., & Geurts, P. A. T. M. (2007). Controversies in water management: Frames and mental models. *Environmental Impact Assessment Review, 27*(7), 685–706.

Lave, T. R., & Lave, L. B. (1991). Public perception of the risks of floods: Implications for communication. *Risk analysis, 11*(2), 255–267.

Linkov I, Wood M. D., Bridges T, Kovacs D., Thorne S., & Butte, G. (2009). Presented at the 2009 IEEE International Conference on Systems, Man, and Cybernetics. San Antonio, TX.

Lowe, T. D., & Lorenzoni, I. (2007). Danger is all around: Eliciting expert perceptions for managing climate change through a mental models approach. *Global Environmental Change, 17*(1), 131–146.

Morgan, M. G., Fischhoff, B., Bostrom, A., & Atman, C. (2002). *Risk communication: A mental models approach*. New York: Cambridge University Press.

National Research Council. (2008). First Report from the NRC Committee on the Review of the Louisiana Coastal Protection and Restoration (LACPR) Program. Washington, DC: The National Academies Press.

Palmgren, C. R., Morgan, M. G., Bruine de Bruin, W., & Keith, D. W. (2004). Initial public perceptions of deep geological and oceanic disposal of carbon dioxide. *Environmental Science & Technology, 38*(24), 6441–6450.

Pielke, R. A., Jr. (1999). Nine fallacies of floods. *Climatic Change, 42*(2), 413–438.

Rabbon, P. D., Zepp, L. J., & Olsen, J. R. (2008). One nation, one policy, one program flood risk management. Presented at the European Conference on Flood Risk Management. Keble College, Oxford, UK.

Rasid, H., & Haider, W. (2002). Floodplain residents' preferences for non-structural flood alleviation measures in the Red River basin, Manitoba, Canada. *Water International, 27*(1), 132–151.

Read, D., Bostrom, A., Morgan, M. G., Fischhoff, B., & Smuts, T. (1994). What do people know about global climate change? 2. Survey studies of educated laypeople. *Risk Analysis, 14*(6), 971–982.

Regnier, E., & Harr, P. A. (2006). A dynamic decision model applied to hurricane landfall. *Weather and Forecasting, 21*, 764–780.

U.S. Army Corps of Engineers, New Orleans District (2008). Louisiana Coastal Protection and Restoration Technical Report Draft.

U.S. Army Engineer Institute for Water Resources. (2009). *Value to the nation: Flood risk management*. Alexandria, VA: U.S. Army Corps of Engineers.

Vari, A., Linnerooth-Bayer, J., & Ferencz, Z. (2003). Stakeholder views on flood risk management in Hungary's Upper Tisza Basin. *Risk Analysis, 23*(3), 585–600.

Wagner, K. (2007). Mental models of flash floods and landslides. *Risk Analysis, 27*(3), 671–682.

Wood, M., Kovacs, D., Bostrom, A., Bridges, T., & Linkov, I. (2012). Flood risk management: U.S. Army Corps of Engineers and layperson perceptions. *Risk Analysis, 32*(8), 1349–1368.

Wood, M., Mukherjee, A., Bridges, T., & Linkov, I. (2009). A mental modeling approach to study decision-making in dynamic task environments. *Construction Stakeholder Management, 240.*

Working Group for Post-Hurricane Planning for the Louisiana Coast. (2006). A New Framework for Planning the Future of Coastal Louisiana after the Hurricanes of 2005. Cambridge, MD: University of Maryland Center for Environmental Science.

Yoe, C. (1993). National Economic Development Procedures Manual: National Economic Development Costs. West Chester, PA: Greely-Polhemus Group, IWR Report 93-R-12.

# Chapter 5
# Adaptive Management for Climate Change

**Matthew D. Wood, Sarah Thorne, Gordon Butte, Igor Linkov, and Daniel Kovacs**

## Introduction

Raising the likelihood of various potential environmental threats such as extreme weather and shifting land use, climate change poses a substantial global risk with many hazards and uncertainties (Stern 2007; Peterson et al. 1997). While discussion of climate change effects is ongoing, large-scale data collection efforts from global air and ocean temperatures, changes in extreme weather event frequency and intensity, and accelerating sea level rise from thermal ocean expansion suggest that some impacts are already being realized and offer some insight into what the consequences

This chapter is adapted from work by Bridges et al. (2013). Climate change risk management: A mental model application. *Environment Systems & Decisions, 33(3),* 376–390. Special thanks to Kelsie Baker and Dr. Todd Bridges for their contributions to this chapter.

M.D. Wood (✉)
U.S. Army Corps of Engineers, Engineer Research and Development Center (ERDC) and Carnegie Mellon University, 696 Virginia Road, Concord, MA 01742, USA
e-mail: Matthew.d.wood@usace.army.mil

S. Thorne, M.A.
Decision Partners, 1084 Queen Street West, #32B, Mississauga, ON, Canada L5H 4K4
e-mail: sthorne@decisionpartners.com

G. Butte
Decision Partners LLC, Suite 200, 313 East Carson Street, Pittsburgh, PA 15217, USA
e-mail: gbutte@decisionpartners.com

I. Linkov
U.S. Army Corps of Engineers, Engineer Research and Development Center (ERDC) and Carnegie Mellon University, 696 Virginia Road, Concord, MA 01742, USA
e-mail: Igor.Linkov@usace.army.mil

D. Kovacs
Decision Partners, 1458 Jersey Street, Lake Milton, OH 44429, USA
e-mail: dkovacs@decisionparters.com



M.D. Wood et al., *Mental Modeling Approach*, Risk, Systems and Decisions,
DOI 10.1007/978-1-4939-6616-5_5

of activities which accelerate climate change may be. The Intergovernmental Panel on Climate Change (IPCC) predicts that global temperatures will increase by 1.1–6.4 °C and sea levels to rise 18–59 cm from 1980 through the end of the twenty-first century (IPPC 2007). Such changes contain immense threats to environmental sustainability alongside drastic threats to human and environmental health, and will significantly challenge those institutions which are responsible for safeguarding the integrity of the environment, the continued usability of coastal installations, the safe and efficient management of water resources, and the protection of a variety of habitats and species which may be permanently altered by a changing environment (Walker and Steffen 1997; Jones 2001; McMichael et al. 2006; Trivedi et al. 2008; Deschanes and Greenstone 2011). To begin to combat this significant environmental concern, government and private sector organizations are collectively adopting predicted impacts of climate change into strategic planning processes (Salazar 2009; Water Utility Climate Alliance 2010; USDOI 2011; USACE 2011; USEPA 2011). However, the high uncertainty posed by climate change makes such preparation difficult at best, where a wide array of potential scenarios and decision alternatives are available for discussion on the subject.

Adaptive management (National Academies 2004) is a robust approach to decision making for management of complex systems (e.g., watersheds). It is ideally suited for a host of natural resource management applications in that it provides a framework for administering systems with many dependencies and stakeholder groups. Most adaptive management processes consist of six key elements. First, management objectives are identified to provide steering principles for which to direct the system. These objectives are regularly revisited and revised based on changes in stakeholder preference/needs and changes in our understanding of the system being managed. Next, a model of the to-be-managed system is developed. This can be a computational model like a simulation; a conceptual model outlined in a flowchart; or another explicit representation that details the bounds of the system, its components, and the affect that each component has on others. Once this model is constructed, a range of management alternatives are identified which may help the system to achieve some desired end state given some current state. Monitoring and evaluation is done to compare the outcomes of management decisions to detect whether and to what extent management objectives are being achieved and the extent to which the management decision is driving the system toward its current state. In order to adapt over time, the adaptive management process needs a mechanism to learn the outcomes of current decisions to enable future decision making. Finally, the process should include a collaborative structure for stakeholders to participate in decision making, e.g., to raise awareness about beneficial or detrimental side effects of management alternatives, and to provide stakeholders with the opportunity to learn about other components of the system as well as the management objectives and interests of other fellow stakeholders. This collection of elements allows system managers and stakeholders to develop robust and flexible management plans that can be modified in the face of unanticipated and changing conditions (Linkov et al. 2006; National Academies 2004).

The United States Army Corps of Engineers (USACE) is an organization which has both military and civil missions to supply engineering support for a wide array of projects and applications. These projects often include coastal navigation infrastructure as a component, making them uniquely exposed to the risks of climate change and requiring USACE to bolster its existing capabilities, assess and prioritize its funding resources, and adapt to potential climate change effects as needed due to its role as a stakeholder in waterway and environmental risk management. Guided by the overarching ideals of environmental sustainability, USACE's general responsibilities include, but are not limited to: constructing and operating navigation infrastructure, supporting commerce and recreation, facilitating storm and hurricane damage reduction, building and managing storm reduction infrastructure such as dikes and levees, reducing flood risks, and protecting and restoring environments in the United States such as its wetlands. In that vein, USACE's Environmental Risk Management efforts seek to identify and manage stressors that interrupt or interfere with its capability to acquire its environmental objectives and outcomes associated with these various civilian and military projects.

To successfully adapt to climate change needs, a review and evaluation of USACE's risk management decision-making process is needed in order to facilitate the agency's adaptation and ability to meet its goals in a shifting and uncertain environment. Where well-informed decision making is a key element in any risk management process, it is essential that USACE's planners develop a comprehensive understanding of likely factors affected by climate change. However, the complexity of this particular decision problem creates difficulties for large organizations such as USACE, which houses a variety of different disciplines, professionals, and policy objectives that operate simultaneously. Additional complications include ever-present concerns of physical, budgetary, and environmental constraints, which can produce conflicting opinions and goals alongside uncertain information that may influence the decision-making process. Despite such uncertainty, these factors must be properly identified, organized, and evaluated in order to facilitate decision making for the particular issue at hand. Falling short of these goals can introduce project oversight, miscommunication, and an inefficient distribution of resources.

Mental Modeling provides a structure to understand and formulate problem-solving strategies and actions for complex domains like climate change adaptation (Morgan et al. 2002; Wood et al. 2012). The approach can be deployed by any public or private organization in a variety of contexts in order to inform comprehensive strategies to address complex decision problems. In essence, Mental Modeling serves as one method to help overcome concerns posed by the uncertainty and share a number of considerations in climate change decision making by integrating various qualitative and quantitative sources of information.

Mental Modeling has been successfully used by several organizations to integrate the understanding of risk into institutional projects and practices (Fischhoff et al. 2011). One of the primary outputs of any mental models exercise is a graphical display that summarizes one or more individuals' beliefs on the variables and factors that influence a decision on a particular topic. Additionally, this display tends to be illustrated as an influence diagram, which uses boxes or ovals (referred to as "nodes") to represent decision variables, and a web of arrows to indicate the relationships between these nodes (Gentner and Stevens 1983; Johnson-Laird 1983; Morgan et al. 1992). The expert mental model is a particular type of mental model that incorporates expert input and is particularly useful to guide agency decision making on complex or sensitive subjects. Expert models offer a formal demonstration of how a particular decision problem or process is understood, integrating the beliefs and opinions of various identified experts in the midst of uncertainty. The ultimate goal of any expert mental model is to provide a summary representation of all relevant decision factors related to the problem of interest (Gentner and Stevens 1983; Johnson-Laird 1983).

Mental Modeling has been deployed to understand and characterize environmental management issues, such as with layperson knowledge of climate change (Bostrom et al. 1994; Read et al. 1994; Reynolds et al. 2010), environmental managers' views on climate change and biodiversity (Hill and Thompson 2006; Lowe and Lorenzoni 2007), public perceptions of carbon sequestration techniques (Palmgren et al. 2004), and laypeople's perceptions of flash floods and landslides (Wagner 2007). USACE is also currently using Mental Modeling to assess flood risk management (Wood et al. 2009), contaminated sentiment management, and beneficial uses of dredged materials (Bridges et al. 2012). Building off these efforts, this paper discusses efforts to characterize the various environmental threats of climate change with respect to USACE's environmental risk management activities by aggregating a collection of internal and external knowledge into a framework in an expert model. The resulting expert models discuss the technical and social issues associated with climate change and were developed within a workshop that incorporated managers and practitioners from a wide array of disciplines within USACE along with relevant external organizations.

## Expert Modeling

In order to acquire much needed information and opinions on the priorities and tasks to improve USACE's environmental risk manage approach for climate change, a series of expert models were developed. These models illustrate an aggregate of the internal and external knowledge and beliefs associated with the relevant scientific, social, and policy issues of climate change on USACE's environmental risk management projects and goals (Decision Partners 2010). Using literature reviews, individual and group interviews, and data analysis, the expert modeling process was initiated with discussions between Decision Partners and experts from USACE to develop a *Base Expert Model of Climate Change Influences on USACE's Environmental Risk Management.* Representatives from USACE's Engineer Research and Development Center (ERDC) then conducted a workshop alongside Decision Partners with 28 experts and stakeholders from the field. Lastly, information acquired from workshop discussion was further analyzed and used to build a series of expert models.

## Results

Looking first at the Base Expert Model, Decision Partners and USACE identified an overview of climate change influences on its environmental risk management practices and beliefs. These variables (represented as nodes in the Base Expert Model of Fig. 5.1) include:

- Climate change drivers.
- USACE environmental risk management process.
- USACE, partner, and stakeholder workforce capacity.
- USACE collaboration, coordination, and communication.
- Uncertainty, variability, and trends.
- Value conflicts regarding resource priority.
- USACE environmental risk management activities.
- Quality of public engagement.
- Individuals' mental models of climate change and USACE environmental risk management.
- Desired outcomes.

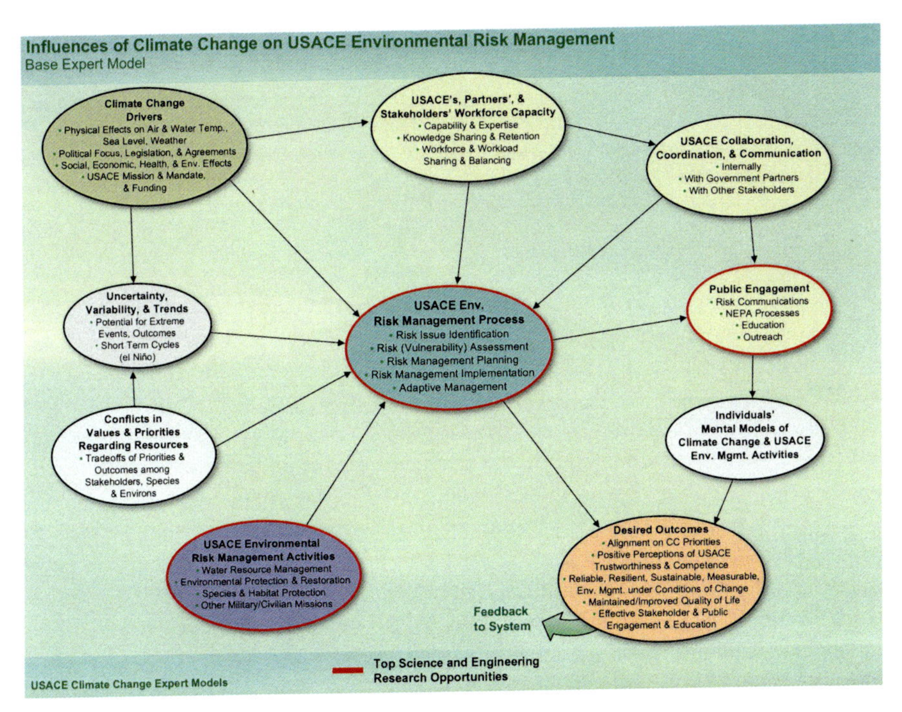


**Fig. 5.1** Base Expert Model showing a high-level overview of the influences of climate change on USACE environmental risk management based on informal interviews with top scientists at the U.S. Army Research and Development Center

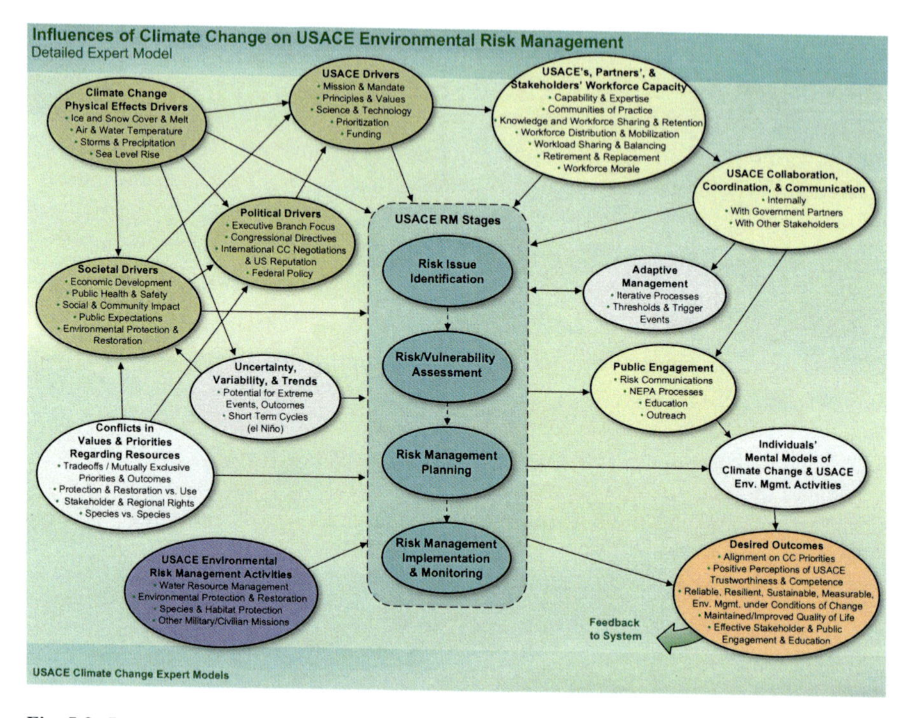


**Fig. 5.2** Detailed expert model showing the influences of climate change on USACE environmental risk management based on the Base Expert Model and input from 28 experts and practitioners from the USACE and external organizations including other federal agencies, academia, environmental consultants, and nongovernment organizations, elicited during the expert model workshop

The 28 workshop participants discussed the earlier variables and added additional components including *Political* and *Societal Drivers*, which facilitated the creation of the Detailed Expert Model. Illustrated in (Fig. 5.2), the Detailed Expert Model is the primary result of the elicitation workshop and maintains a similar structure as the Base Expert Model. This figure is best read by beginning in the top left corner, then following nodes in the middle of the figure, and then finally reaching *Desired Outcomes* regarding the Model's established goals.

The workshop participants particularly indicated that an opportunity exists to better understand and discuss the consequences of climate change to the built environment and undeveloped areas alike. Specifically, these physical effects are aggregated in the *Climate Change Physic Effects Drivers* node, and can be utilized as inputs in risk management planning processes and exercises, and effectively allow for more accurate indicators in particular areas such as with *Ice and Snow Cover & Melt* or *Storm and Precipitation Changes*. Workshop participants also noted that such climate change risk management processes should be developed for individual regions and species, where drivers will have unique effects and circumstances in each individual case.

Next, the workshop participants outlined various *Political Drivers*, which generally compromise those influences from outside agencies, organizations, and stakeholders that have an impact on organizational priorities. Relevant variables here include *Executive Branch Focus* (on climate change issues), *Congressional Directives*, *International Climate Change Negotiations*, and *Federal Policy*. Overall, the workshop participants argued that there is a substantial need to improve understanding of relevant governance challenges related to climate change. Where climate change is a complex issue on both a political and environmental scale, the workshop participants noted that existing regulations or policies may be insufficient to address emerging climate change concerns.

Societal Drivers are defined as those issues where society at large has a vested interest and can potentially drive USACE environmental risk management activities in order to reduce the negative environmental health hazards to society. Relevant variables in this node include *Economic Development* (protect against economic damage), *Public Health and Safety*, and *Environmental Protection and Restoration.* Workshop participants discussed the increase in climate change awareness over the years, although this awareness has not directly contributed to specific action on environmental policy. This is potentially due to a lack of clarity regarding what the public actually wants, where some question the need to act at all while others demand substantial changes in environmental policy and reduced tolerance for environmental pollutants.

Looking next at USACE Internal Drivers, workshop participants defined these as originating within USACE proper and have some influence on their environmental risk management policies and activities. Relevant variables here include *Mission and Mandate, Principles and Values, Science and Technology, Prioritization,* and *Funding.* These collective variables include an overall focus in public protection, water resource management, environmental protection, waterway infrastructure, warfighting, and homeland security. Of these, workshop participants noted the particular need to invest more funding in measurement and monitoring activities in order to support adaptive management activities.

Next, the central node represents the *USACE Environmental Risk Management* process. The key stages here include risk issue identification, vulnerability assessment, risk management planning, and risk management implementation and monitoring. It should be noted that adaptive management should be closely integrated with the environmental risk process as well, which is formally illustrated with the double-headed arrow between *Adaptive Management* and *USACE RM Stages* nodes.

The bottom left corner node in the Detailed Expert Model includes those factors involved in USACE Environmental Risk Management Activities. USACE roles here include *Water Resource Management, Environmental Protection and Restoration, Species and Habitat Protection,* and *Other Military and Civilian Missions.* For those concerns specific to USACE's environmental risk management activities, workshop participants noted a need to better understand the impacts of climate change on water resource management and the habitats of various vulnerable species. Table 5.1 presents the full list of specific recommendations for USACE environmental risk management activities.

**Table 5.1** Opportunities relevant to the primary roles of the USACE in environmental risk management, as identified during the expert elicitation workshop

| Environmental risk management activity | Needs and opportunities |
|---|---|
| *Water resource management* | • Reduce the impact of climate change on water supply, flood protection, waterway navigation, and waterway recreation, through management of water resources such as reservoirs and waterways<br>• Improve USACE infrastructure activities in support of water resource management endpoints<br>• Better understand the issues and opportunities related to USACE water resource management under conditions of climate change |
| *Environmental protection and ecosystem restoration* | • Improve support for overall ecosystem health in many watersheds<br>• Better understand challenges to ecosystem restoration<br>• Adopt lessons learned from ecosystem management activities at Eglin Air Force Base |
| *Species and habitat protection* | • Improve management of species vulnerability, which is a central issue with respect to climate change impacts<br>• Better understand and characterize the impacts of USACE activities on species<br>• Better understand the complexity of species restoration in the context of USACE environmental management activities<br>• Identify variables for performing vulnerability assessments<br>• Develop decision analyses for protection of threatened and endangered species and incorporating these into USACE planning activities |
| *Military missions* | • Better understand and characterize the potential impacts of climate change on Army military installations and missions in order to maintain facility function, meet training needs, and adapt operations |

The Model endpoint centers on *Desired Outcomes*, which is the culmination of USACE goals for climate change regarding the organization's risk management processes. Among others, these goals include *Alignment of Climate Change Priorities, Positive Perceptions of USACE Trustworthiness, Reliable—Resilient—Sustainable and Measureable Environmental Management,* and *Effective Stakeholder and Public Engagement and Education.* Overall, the *Desired Outcomes* node is heavily influenced by *USACE Risk Management Stages* and *Individuals' Mental Models of Climate Change*.

## Conclusions

This case presented a demonstration of how USACE applied Mental Modeling techniques to develop Adaptive Management strategies to better adapt to changes to its business as a result of climate change. Mental Modeling in this case helped to establish a model of the system that could then be used to inspire inquiry into management alternatives, learning mechanisms, stakeholder engagement opportunities, and other essential elements of an adaptive management process. USACE, together with its partners and stakeholders, concluded that opportunities were available for improving environmental risk management practices to more adequately address climate change impacts using an adaptive management framework in a way that is consistent with stakeholder values. Indeed, values are represented in two nodes of the Detailed Expert Model: *USACE Drivers* and *Conflicts in Values and Priorities Regarding Resources* (Fig. 5.2). Both in turn influence how people respond to climate change threats. Future work could identify synergies and antagonisms between USACE values and social–political values to develop a more robust understanding of the relationship between personal values and the perceptions and reactions to climate change risks.

Conclusions reached by USACE and its partners through the Mental Modeling process are consistent with conclusions of other federal and international agencies. For example, U.S. Fish and Wildlife Service (FWS) developed their Climate Change Strategic Plan, emphasizing increased coordination and collaboration with other agencies and organizations, flexible, adaptable management techniques, use of state-of-the-art technology, improvement of modeling and monitoring efforts, and involvement of stakeholders and the public through effective information communication (FWS 2010). The Canadian Communities' Guidebook for Adaptation to Climate Change similarly emphasizes communication and collaboration, flexibility, monitoring, and utilization of the latest information (Bizikova et al. 2008).

Potential climate change impacts are a significant challenge to agencies charged with managing environmental risks because of the uncertainty associated with potential impacts. To better prepare, the USACE intends to update environmental risk management (ERM) practices to incorporate predicted influences of climate change, in part through evaluating the factors affecting ERM decisions based on relevant knowledge of experts from inside and outside of USACE. Climate change presents significant decision-making challenges because of the complexity of climate change influences on environmental risk management. The expert models presented here can help to establish a common reference frame to reduce some of that complexity and enable understanding among those making or affected by ERM decisions and inform USACE strategies for improving ERM in the context of climate change.

## References

Bizikova, L., Neale, T. L., & Burton, I. (2008). *Canadian communities' guidebook for adaptation to climate change: including an approach to generate mitigation co-benefits in the context of sustainable development.* Adaptation and Impacts Research Division, Environment Canada.

Bostrom, A. M., Morgan, G., Fischhoff, B., & Read, D. (1994). What do people know about global climate change? 1. Mental models. *Risk Analysis, 14*(6), 959–979.

Bridges, T. S., Kovacs, D., Wood, M. D., Baker, K., Butte, G., Thorne, S., & Linkov, I. (2013). Climate change risk management: A mental model application. *Environment Systems & Decisions, 33*(3), 376–390.

Bridges, T., Lu, C., Wood, M. D., Kovacs, D., Thorne, S., Butte, G., Wilson, J., & Linkov, I. (2012). *Increased environmental beneficial use of dredged material: Application of expert mental modeling*. Vicksburg, MS: U.S. Army Corps of Engineers Technical Note.

Decision Partners, L. L. C. (2010). *Influences of climate change on environmental risk management: Expert model narrative*. Pittsburgh: Decision Partners.

Deschanes, O., & Greenstone, M. (2011). Climate change, mortality, and adaptation: Evidence from annual fluctuations in weather in the US. *American Economics Journal: Applied Economics, 3*(4), 152–185.

Fischhoff, B., Neuhauser, L., Paul, K., Brewer, N. T., & Downs, J. (2011). *Communicating risks and benefits: An evidence-based user's guide*. Silver Spring, MD: Food and Drug Administration.

Gentner, D., & Stevens, A. L. (1983). *Mental models*. Hillsdale, NJ: Erlbaum.

Hill, S. D., & Thompson, D. (2006). Understanding managers' views of global environmental risk. *Environmental Management, 37*(6), 773–787.

IPPC. (2007). Climate Change Synthesis Report. Contribution of Working Groups I, II and III to the Fourth Assessment Report of the Intergovernmental Panel on Climate Change.

Johnson-Laird, P. N. (1983). *Mental models: Towards a cognitive science of language, inference, and consciousness (No. 6)*. Cambridge, MA, USA: Harvard University Press.

Jones, R. N. (2001). An environmental risk assessment/management framework for climate change impact assessments. *Natural Hazards, 23*(2-3), 197–230.

Linkov, I., Satterstrom, F. K., Kiker, G., Batchelor, C., Bridges, T., & Ferguson, E. (2006). From comparative risk assessment to multi-criteria decision analysis and adaptive management: Recent developments and applications. *Environment International, 32*(8), 1072–1093.

Lowe, T. D., & Lorenzoni, I. (2007). Danger is all around: Eliciting expert perceptions for managing climate change through a mental models approach. *Global Environmental Change, 17*(1), 131–146.

McMichael, A. J., Woodruff, R. E., & Hales, S. (2006). Climate change and human health: Present and future risks. *The Lancet, 367*(9513), 859–869.

Morgan and Henrion (1990) should be Morgan, Henrion, and Small (1992) and is in the reference list.

Morgan, M. G., Henrion, M., & Small, M. (1992). *Uncertainty: A guide to dealing with uncertainty in quantitative risk and policy analysis*. Cambridge, NY: Cambridge University Press.

Morgan, M. G., Fischhoff, B., Bostrom, A., & Atman, C. (2002). *Risk communication: A mental models approach*. Cambridge, NY: Cambridge University Press.

National Academies. (2004). *Adaptive management for water resources project planning*. Washington, DC: The National Academies Press.

Palmgren, C., Morgan, C., de Bruin, W., & Keith, D. (2004). Initial public perceptions of deep geological and oceanic disposal of carbon dioxide. *Environmental Science and Technology, 38*(24), 6441–6450.

Peterson, G., De Leo, G. A., Hellmann, J. J., Janssen, M. A., Kinzig, A., Malcolm, J. R., & Tinch, R. R. (1997). Uncertainty, climate change, and adaptive management. *Conservation Ecology, 1*(2), 4.

Read, D., Bostrom, A., & Smuts, T. (1994). What do people know about global climate change? Part 2: Survey studies of educated laypeople. *Risk Analysis, 14*(6), 971–982.

Reynolds, T., Bostrom, A., Read, D., & Morgan, M. (2010). Now what do people know about global climate change? Survey studies of educated laypeople. *Risk Analysis, 30*(10), 1520–1538.

Salazar, K. (2009). Secretarial order no. 3289: Addressing the impacts of climate change on America's water, land, and other natural and cultural resources. Washington, DC.

Stern, N. (2007). *The economics of climate change: The Stern review*. Cambridge, NY: Cambridge University Press.

Trivedi, M. R., Berry, P. M., Morecroft, M. D., & Dawson, T. P. (2008). Spatial scale affects bioclimate model projections of climate change impacts on mountain plants. *Global Change Biology, 14*(5), 1089–1103.

U.S. Army Corps of Engineers (USACE). (2011) USACE Climate change adaptation policy statement. www.usace.army.mil/environment/Documents/USACEClimateChangeAdaptationPolicy Jun11.pdf.

U.S. Environmental Protection Agency (USEPA). (2011). Policy statement on climate-change adaptation. http://www.epa.gov/climatechange/effects/downloads/adaptation-statement.pdf.

U.S. Fish and Wildlife Service (FWS). (2010). Rising to the urgent challenge: Strategic plan for responding to accelerating climate change. http://www.fws.gov/home/climatechange/pdf/CCStrategicPlan.pdf.

Wagner, K. (2007). Mental models of flash floods and landslides. *Risk Analysis, 27*(3), 671–682.

Walker, B., & Steffen, W. (1997). An overview of the implications of global change for natural and managed terrestrial ecosystems. *Conservation Ecology, 1*(2), 2.

Water Utility Climate Alliance (WUCA). (2010). Decision support methods: Incorporating climate change uncertainties in water planning. www.wucaonline.org/assets/pdf/pubs_whitepaper_012110.pdf.

Wood, M., Mukherjee, A., Bridges, T., & Linkov, I. (2009). A Mental modeling approach to study decision-making in dynamic task environments. *Construction Stakeholder Management, 240.*

Wood, M., Kovacs, D., Bostrom, A., Bridges, T., & Linkov, I. (2012). Flood risk management: U.S. Army Corps of Engineers and layperson perceptions. *Risk Analysis, 32*(8), 1349–1368.

# Chapter 6
# Technology Infusion and Marketing

**Matthew D. Wood, Sarah Thorne, and Gordon Butte**

## The Opportunity

The EL invents. Individual investigators and teams routinely solve problems presented by customers of different types in the Civil Works and Military missions supported by the USACE. Solutions may involve the creation of new products and services, as well as new capabilities—methods and technologies. Development efforts have also resulted in an extensive portfolio of solutions that have the potential to add value in a number of markets and for customers in addition to those for whom the solutions were initially developed.

The opportunity is for the EL to also innovate not only in science and technology product development, but also in TIM. Here innovation is defined as the process of assuring the successful use of an invention—a solution to a problem—among given customers that comprise markets.

Dr. Elizabeth Fleming, head of the ERDC Environmental Lab has noted that, in order to accomplish innovation the EL has to satisfy its own customer-related needs. In a commentary on the challenge of Technology Infusion and Marketing, Dr. Fleming said: "We need EL customers to, first, be aware of the creative solutions that we have developed and can develop and, second, to understand the utility and value of our solutions and, third, to know from us and with our support how best to

M.D. Wood (✉)
U.S. Army Corps of Engineers, Engineer Research and Development Center (ERDC) and Carnegie Mellon University, 696 Virginia Road, Concord, MA 01742, USA
e-mail: Matthew.d.wood@usace.army.mil

S. Thorne, M.A.
Decision Partners, 1084 Queen Street West, #32B, Mississauga, ON, Canada L5H 4K4
e-mail: sthorne@decisionpartners.com

G. Butte
Decision Partners LLC, Suite 200, 313 East Carson Street, Pittsburgh, PA 15217, USA
e-mail: gbutte@decisionpartners.com



M.D. Wood et al., *Mental Modeling Approach*, Risk, Systems and Decisions,
DOI 10.1007/978-1-4939-6616-5_6

apply the solutions to realize their full value. We also need to demonstrate the value of our programs by better articulating what we can contribute and how we make a difference in Civil Works and Military programs and missions. Our diverse portfolio of science and technology solutions allows us to leverage our Civil Works and Military solutions by transferring existing technology to new customers and new technologies to existing customers. And we need to take risks in a smart manner. Nothing will change for the better until we do. The leadership of the EL is fully committed to enabling us all to do so."

EL, as a part of the USACE more broadly, has both a Civil Works and a Military mission, each of which are associated with different business lines and distinct though often overlapping customer bases. The Civil Works mission is to "[provide] quality, responsive service to the nation in peace and war (USACE 2014a)." These activities include infrastructure development and support, flood risk management, environmental stewardship, and others. USACE's Military mission is primarily focused on "[providing] premier engineering, construction, real estate, stability operations, and environmental management products and services for the Army, Air Force, other assigned US Government agencies and foreign governments (USACE 2014b)." This includes such activities as infrastructure support as well as research and development for the U.S. Army and other U.S. Armed Services domains.

Many prominent researchers, business growth consulting professionals, and investors (e.g., Bacon and Butler 1973, 1998) have noted that key to technology infusion success is matching a given product to the customer's greatest need and communicating during development, ensuring the product generates full value for the customer. While this is well recognized by marketers, it is often unclear to a product developer. How to accomplish this matching of need and value and do so fast, accurately, cost effectively, and efficiently will significantly reduce the risk of mismatch leading to product market failure.

Research on technology transfer for federal labs identified disagreements on both methods and terminology. No single approach was identified as applicable to EL and ERDC. That said, past research conducted for the ERDC highlighted a few specific factors to be considered for effective TIM, including: Deliberate approaches to TIM, valuing successful TIM within the organization; providing funding for TIM activities; having organizational leadership promote TIM as a priority; and cultivating champions within the organization to promote TIM (Decision Partners 2011). This work, combined with a review of successful corporate innovation processes, suggested the Mental Modeling approach, if tailored to fit the circumstances and business culture of EL, could be used to improve TIM in the organization (Fig. 6.1).

The hypothesized approach focused on three key steps:

1. Characterizing the TIM opportunity for EL technologies, products, and services;
2. Validating these hypotheses of opportunity; and
3. Implementing a Marketing Plan to achieve these opportunities.

Workshop participants focused the first step, Opportunity Characterization, for three EL technologies. Using templates the breakout groups defined:

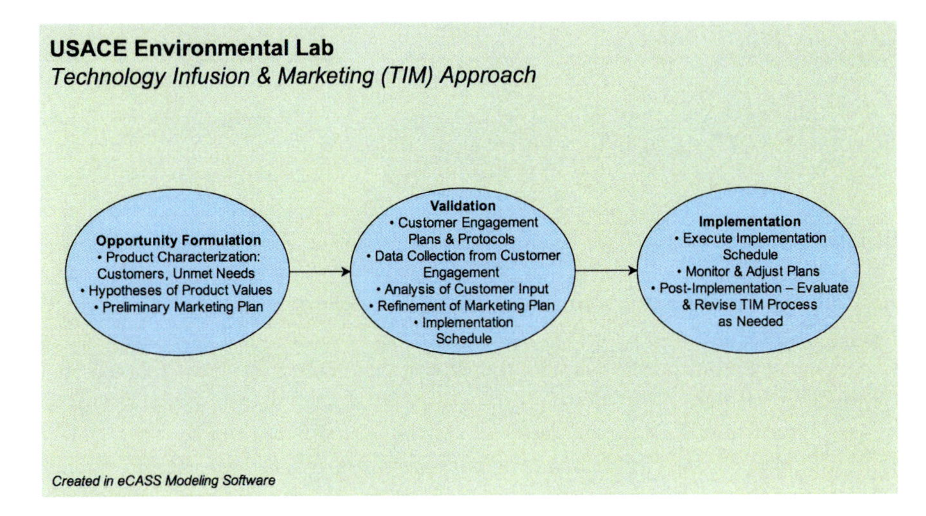


**Fig. 6.1** Base model of the TIM approach

- Comprehensive product descriptions.
- Potential customers and their unmet needs.
- Hypotheses of product values to potential customers.
- Marketing plans and immediate next steps in preparation for Validation (Step Two).

## Base Model of the TIM Approach

The TIM approach focuses on identifying detailed hypotheses of customer need and value for a given product with a given customer that can be tested directly with customers using a mental models-type inquiry. Test results are used to inform key product development and marketing decisions, including modification of products to closely meet customer needs and deploying marketing communications specifically tailored to customer mental models and purchase decision making.

## Step One: Opportunity Formulation

The purpose of the first step, Opportunity Formulation, is to provide a simple structure, or model, to guide innovators in assembling a mental model of the product market opportunity. This is done through a process of identifying and thinking through the components of a product, service, or capability that, in various combinations, could represent a solution to a given customer's problem.

## *Critical Definitions*

A Solution could be a product or a service/capability, or in some cases, a product supported by service/capability, defined as the following.

### Product

- A product is a hardware item, software, or process that a customer can acquire and implement directly.
- The key distinguisher between a product and a capability/service is that the customer can acquire or access the former directly, with little or minimal technical assistance from ERDC. *If you can't provide a distribution source for the technology, it probably isn't a product. If a distribution source will be available, but not until sometime in the future, chances are the topic is "ongoing research."*
- Some products may be commercially available, others distributed through ERDC websites, and some may even be the website itself when the concept and operation was developed by ERDC. Examples of the latter: NRM Gateway, DENIX.

### Service/Capability

- A service or capability is technology support provided to a customer, usually on a reimbursable basis.
- A capability could include providing a customer with access to unique experience and expertise of ERDC personnel or access to ERDC facilities to conduct research, experiments, or tests to solve customer problems. It could mean experts apply a product (the "supporting technology") to solve a problem on behalf of the "customer." For a direct-funded program, the "customer" could be a USACE directorate and the Assistant Secretary of the Army for Acquisition, Logistics, and Technology (ASA(ALT)).
- A service could include the customized application of an ERDC developed technology and the implementation of that customized technology solution for a particular customer.

The basic process for Opportunity Formulation comprises three tasks each executed at the appropriate level of detail: (1) characterize the product or service in-depth; (2) formulate detailed hypotheses of key or strategic customers and their unmet needs, solutions of highest value to the customer that fit with their unmet needs, and identify the full value of the solutions in qualitative and quantitative terms; and (3) prepare a preliminary marketing plan comprised of detailed outlines for engagement with customers, partners, and suppliers, leveraging existing relationships, if any.

To guide deliberation in the tasks, key questions to be addressed include:

1. What is the product, described as generically as possible?
2. Who are the customers—current, potential, and possible customers—and where are they located?
3. (For each customer) What performance requirements or needs do the customers have that the product can deliver?
4. How can the product value be communicated in an appealing way? What will the customer buy? What is the hypothesis of a solution to the customer's unmet needs for which they are willing to pay?
5. What is the situation today? How is the customer's problem being addressed now, if at all?
6. What are the full costs of how the problem is being addressed now?
7. What is wrong with the way it is done today? What improvements could be made?
8. What might the value of those improvements be?

## Step Two: Validation

Opportunity Formulation calls for development of hypotheses of the product or service, a solution to a customer problem, and the value of that solution. Consistent with the scientific method, hypotheses are considered to be neither right or wrong and must be tested, or validated, to determine what is right or wrong and why those factors are right or wrong.

The key tasks in the validation process for TIM are as follows: (1) prepare customer engagement plans and protocols, with an emphasis on existing customers where possible; (2) engage customers in dialogue using the protocols to systematically collect data for understanding each hypothesis in depth; (3) analyze data to identify the total requirements for success of a given solution, by customer, and why they are the requirements; (4) refine the marketing plan based on this analysis; and (5) prepare an implementation schedule.

## Step Three: Implementation

With the requisite insights and data from the validation activity in Step Two, the implementation tasks can be completed with confidence. These tasks include: (1) execute the implementation schedule, (2) frequently monitor and adjust the implementation process as required, (3) evaluate postimplementation the activities in all steps and their results to identify TIM process improvements, and (4) revise the TIM process as warranted.

## Workshop: Technology Infusion and Marketing (TIM): Guided Thinking on Three Technologies

EL leaders organized a 2-day, professionally facilitated workshop for 30 scientists and project leaders to jointly prepare detailed product market opportunity and value descriptions as well as an outline business model for three technologies available at the EL. EL leadership wanted to enhance the ability of EL personnel to transfer and market products by effectively communicating how those products solve the customer's problems. The purpose of the workshop was to identify how to improve performance in technology infusion and marketing, technical competence, and knowledge management by the EL. The planned outcome: comprehensive business models and plans for three EL products or services/capabilities using EL's state-of-the-science TIM Approach.

The three products identified for participants to focus on were the following:

- TREECS—Training Range Environmental Evaluation and Characterization System—characterized as both product and service/capability.
- Computational Chemistry—characterized as service/capability.
- Risk Management—characterized as a service/capability.

## Preworkshop with EL Sponsors and Project Leads

A preworkshop meeting with sponsors and project leads from EL was conducted by the facilitators to better understand and articulate the opportunity for TIM for specific projects or services/capabilities currently under development by EL that had high potential for technology transfer. In the meeting, several opportunities were identified to increase the likelihood of adoption of the product or service/capability solutions developed by EL. A priority was to find ways to make current and new EL customers more aware of the solutions that EL develops, the value or utility of these solutions, and the ways that these solutions might be applied to the unique challenges faced by these customers. These outreach and awareness building efforts could be tailored to customers served by Civil Works and Military business lines, which have distinct, but often overlapping needs. For both of these customer groups, means were identified to transfer existing technology to new customers and new technologies to existing customers. Finally, participants in the preworkshop meeting noted that there was an opportunity to promote calculated risk taking in business development initiatives in order to improve current TIM practices.

Four broad workshop objectives were identified by EL leaders and stated as: Building on recent research in technology and marketing in EL, ERDC, and USACE in general in order to:

1. Build shared understanding of the context and challenge of improving TIM across EL;

2. Develop comprehensive business models for three EL technologies in separate breakout groups, and refine these models based on shared plenary input;
3. Develop outline marketing plans for each technology and refine them based on plenary input; and
4. Identify and discuss immediate next steps, including validation.

In addition to these objectives, EL leaders wanted to use the workshop to develop a model approach for TIM that could be used across EL, and possibly by other laboratories at ERDC along with federal laboratories outside USACE.

## Facilitators' Protocol

Information from preworkshop meetings and concepts identified in secondary research were used to develop a structured protocol for conducting the workshop in order to most effectively understand current TIM successes and develop a business process to increase success in the future. A detailed facilitator's protocol was developed to help facilitators focus workshop participants in their efforts to complete each of the tasks in each step of the process, answering the guiding questions presented in the tasks to do so.

## Workshop Agenda Overview

Day One started with a plenary with EL leaders providing opening comments and background to help participants develop a shared understanding of the challenges currently faced in improving TIM. Recent successes in implementing TIM within EL were presented in the form of case studies on three very different EL products and services.

Participants were then organized into three breakout groups which worked concurrently to begin work on development of their TIM Business Model of one of the three topics: TREECS, Computational Chemistry, and Risk Management. In facilitated breakout groups, the teams worked through the questions and summarized their discussions on worksheets. At the end of the day, each group presented a progress report to the plenary.

Day Two began with a discussion on the TIM business model, followed by the breakout groups continuing work on their product or service. In the second breakout, they refined their respective TIM models based on plenary feedback, and then worked on the business and marketing plans for their product or service/capability. Day Two concluded with a plenary where each breakout group presented their revised TIM business models, solicited feedback, then worked together to refine the TIM business model so it would be applicable across a broad range of current and future EL & ERDC technologies. Senior EL leaders attended the final plenary and provided comments and insight in how to take the three business opportunities to the next step—Validation.

## Breakout Group Results

Each breakout group focused their efforts on one of the products or services described earlier. They began with a project description that was quite narrow in scope. Through discussion guided by the TIM approach, the groups were able to develop a more comprehensive conception of the product or service/capability that was complete and robust enough to advance to the next step, Validation. For each breakout group, a brief summary is given below of the product or service/capability description, the identified customers and their unmet needs, the hypotheses of product value, and the marketing plan with next steps identified in the breakout group.

## TREECS: Training Range Environmental Evaluation and Characterization System

TREECS is a client-based software system that provides rapid assessment of the transport, fate, and risk to soil constituents of contamination. For more complicated applications, a consulting service can be incorporated to help customers develop an informed land/site management plan utilizing the modeling results of the TREECS software. It is currently optimized for military ranges to assess how munitions materials travel through the soil, but can support other customers and needs as well. For instance, TREECS could serve other military and nonmilitary customers with a need to characterize fate and transport of contamination through a terrestrial environment over time. The identified hypothesis of product value for TREECS was that it "enables rapid, simple, cost-effective, predictive assessment of sites to enable proactive site management reducing the need for extensive physical sampling and, reducing or preventing costly environmental mitigation activities and potential loss of site use."

The TREECS marketing plan and next steps identified in the workshop included several initiatives. Among them were investigating ways to incorporate the TREECS approach and process into Army standards. This would help Army to project, at the research or acquisition phase, the impacts of materials they are developing and purchasing, and may help to reduce problems with site contamination, for example, further along in the product life cycle. The TREECS group noted that continued communication with and education of potential customers and partners was particularly important and could include online portals such as the TREECS website and webinars, plus one-on-one meetings and workshops. The group identified future development opportunities, such as incorporating cost functions into TREECS that could highlight the cost savings potential from proactive site management, as well as renaming the software system so that it reflects a broader application potential, e.g., *Terrestrial Rapid Environmental Evaluation and Characterization System*, which was subsequently done.

## Computational Chemistry

Computational chemistry is a service/capability that uses software simulations of materials commonly used in industrial processes to better understand their characteristics and potential interactions with other materials. Customers for the service are mid- to upper-level scientists in the military (e.g., munitions manufacturing) and private sectors (e.g., pharmaceutical industry, municipalities) with a need to screen the potential impact of new or proposed materials and their degradation products, including impacts on the environment. In particular, Computational Chemistry targets the Army Acquisition community. The service/capability could have an impact on this community early in their decision making on new materials, which could include a wide variety of applications including those used in munitions as well as new fabrics for soldier clothing. The identified hypothesis of product value for Computational Chemistry was that "cost-effective screening of new materials could enable a maximization of investments by minimizing potential environmental, health, and safety impacts and resulting financial and regulatory liability that could result from a loss of production/utility and/or significant environmental impacts over time."

The marketing plan and next steps identified in the workshop for Computational Chemistry included four main points. First EL and senior Army leaders should meet to discuss capabilities at ERDC in an effort to open communication with technical-level decision makers. EL leadership can also be instrumental in facilitating collaboration and internal alignment within ERDC. Members of the Computational Chemistry team should meet with key individuals who can influence strategic customers and validate the hypothesis of value. The team should also conduct demonstration case studies and include these cases in marketing materials.

## Risk

Risk is a service/capability that uses basic and applied research and tools to characterize risk in many contexts and enable sustained development and continued operation by providing adequate knowledge to inform leaders and decision-makers. Customers for this service are military and government agencies who have an immediate need for science-based risk evaluation and characterization, and clear comprehensive guidance to support decision making. The identified hypothesis of product value for Risk was that "EL has highly capable personnel and high quality facilities that many potential customers do not have access to, supplying them with cost-effective risk characterization and management guidance that allows for effective decision making and solution of customer problems that save customers significant time and money from fieldwork, rework and delayed action on risk-based decisions."

The marketing plan and next steps identified in the workshop for Risk included a few initiatives. One such initiative was to network and seek potential customers and deliver a short overview describing the package of services that could be performed to meet the customer's needs. In addition, communication materials, such as fact sheets and success stories with contact information, should be developed to provide potential customers with a summary of the risk capability, what it can do for them, and whom to contact for more information. A concrete engagement plan should also be developed to make initial contact and stay engaged with potential customers.

## Key Learnings and Applying the Results

The results of the workshop successfully validated the general TIM approach for application to specific EL technologies, products, and services/capabilities. As applied to EL specifically, the approach proved to be an action-oriented, results-producing process for achieving TIM through better understanding of EL technology, product, and service/capability opportunities. It provided a structured approach for determining the total requirements that provides the customer with success through a specific technology, product, or service/capability. This approach incorporates three steps, Opportunity Formulation → Validation → Implementation (Fig. 6.1) in a hypothesis-driven fashion to predict performance and customer success based on in-depth understanding of customer unmet needs and the total requirements of success. It uses clear procedures for each key task in each step in order to systematically and repeatedly produces the highest value solution to address the customer's unmet needs.

The Expert Model of the TIM business model for EL technologies, products, and services/capabilities specifically can be found in Fig. 6.2 and incorporates common considerations identified across the developed TIM business opportunities for each of the three EL technologies. The expert model is made up of nodes that are drivers or key variables to be considered when developing and engaging a new TIM effort, and arrows that represent the direction of influence between these variables. The model is read from top-left to bottom-right, starting with internal EL drivers of the TIM process (tan), incorporating project-centric variables (purple), external influences (yellow), the USACE EL Technology/Product/Service Opportunity of interest (blue), which is influenced by the other drivers above, and finally Outcomes (orange) that are desired from successful implementation of a TIM business model.

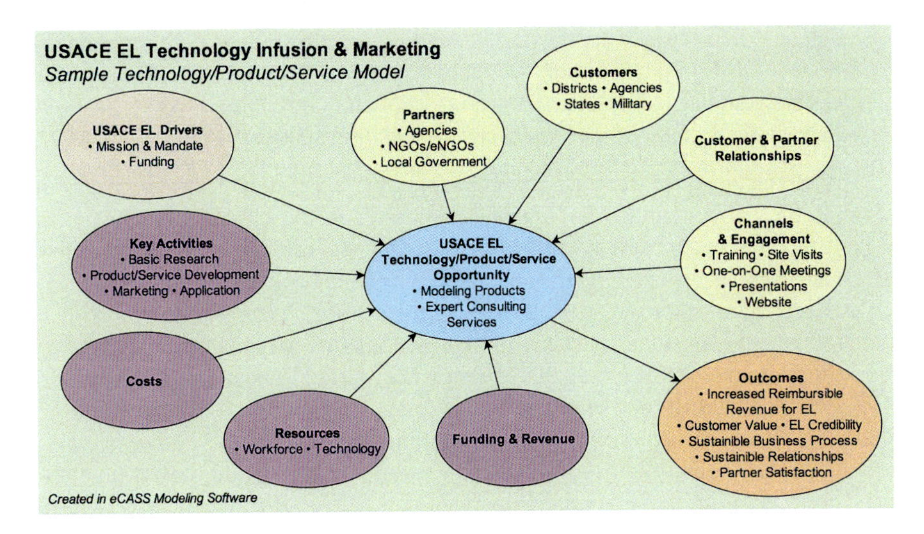


**Fig. 6.2** Expert model of technology infusion and marketing business model

## TIM Path Forward Considerations/Action Options

The TIM workshop generated a number of key considerations, or action options, for advancing TIM at the Environmental Lab (EL), throughout ERDC and beyond. Considerations included the following:

1. Confirm the Opportunity Validation and Business Plan model as a task-suited approach for accomplishing TIM at the EL. A general business model and business plan approach was developed for the workshop and validated through application by participants and EL executive. The general approach was derived from a review of relevant literature on technology transfer in a federal lab (and university lab) setting (e.g., Bozeman 2000), and relevant literature and experience for innovation through product/service development and marketing in a corporate setting.[1] The validated approach or model can be confirmed by EL executives.
2. Conduct workshop(s) for three or more TIM products. Using the validated approach (Opportunity Validation and Business Plan model), workshops can be held for addressing one or more sets of three TIM products. The Opportunity Validation and Business Plan model and workshop design can be further refined or customized to application in ERDC and the USACE with workshop results.
3. Migrate the Opportunity Validation and Business Plan model to other labs and organizations within ERDC that want to improve their TIM performance.
4. Build internal capacity for the application of the Opportunity Validation and Business Plan model. The EL has created leading practice in TIM for federal labs and government organizations with a TIM-relevant mission of priority. The

[1] In particular, Planned Innovation and experience at the Dow Chemical Company. See: http://plannedinnovationinstitute.com/.

EL is now in a position to transfer the leading practice to many other organizations as a TIM initiative or product generated and proven at the EL. The transfer initiative could be led by a team of expert facilitators from the EL who have been trained in facilitation of the Opportunity Validation and Business Plan model.

These considerations are being implemented to various degrees. The goal is to increase the Environmental Laboratory's relevance to the USACE and Army, as well as its other customers, by expanding the reach and use of its products and services/capabilities through effective Technology Infusion and Marketing (TIM).

In April 2014, approximately thirty scientists from across the Environmental Laboratory came together for the second Technology Infusion and Marketing (TIM) workshop sponsored by Dr. Beth Fleming. The participants represented diverse backgrounds, interests, and expertise and a range of experience. The purpose of the Workshop was to identify new, bold, cross-cutting, and integrated Solutions to environmental science and engineering challenges and to develop concept plans (hypotheses) for marketing those Solutions to customers. EL's TIM process was used to generate and prioritize Solutions to address the challenges described in each of the following categories:

1. Military—Toxic Byproducts of Synthetic Biology.
2. Military/Civil Works—Unmanned Aircraft Systems for Environmental Data Capture.
3. Civil Works—Integrated Water Resources Technology.
4. Civil Works—Planning and Decision Support Mechanisms.

Problem Statements for each category were developed to focus the groups on challenges resulting from the USACE, Army, and Department of Defense (DoD) strategic directions described in the *Civil Works Research and Development Strategy, 2014*; the *Reliance Operating Principles, 2014*; *Science and Technology Trends, 2013–2043*; *A Review of Leading Forecasts, 2014*; and *The Quadrennial Defense Review, 2014*.

## An Example of Project-Specific Successes and Learnings: TREECS

A follow-up meeting was conducted with EL TIM project sponsors in April 2014 to discuss successes from the initial TIM workshop initiative and to prepare for a second TIM workshop with a new set of EL projects. The sponsors reflected positively on the results provided by the TIM business model developed for the different projects, particularly for TREECS. The TIM model for TREECS helped EL to expand the customer base well beyond the training range context for which it was developed. Consideration of a larger customer base helped EL to rebrand TREECS by making a slight change to the name, which paid large dividends. The researchers noted, "...we got out and started talking about TREECS, the Terrestrial Rapid

Environmental Evaluation and Characterization System, and all of a sudden, we have customers coming out of the woodwork." These included a large number of customers who were interested in analysis of material fate and transport, just like the training ranges that were instrumental in developing the capability, but had quite different interests in application. Some intelligence and enforcement agencies were interested in using these analyses as a way to monitor potential adversarial activity and in detecting the manufacture of illegal drugs. DoD agencies responsible for managing facilities were interested in using the toolkit to identify the responsible party for cleanup of perchlorate on DoD-owned facilities that were leased to private interests to provide mission support. The result of this rebranding and expanded interest from new customers was over $3 million in new business over a period of 18 months.

The TIM business model helped EL to realize that their marketing efforts for TREECS had been limited by their imagination in selling the capability and not by the ability of the tool to apply to these new domains. The project sponsors also realized that TREECS helped EL to develop a lot of expertise that could be used to expand EL's reach in the types of environmental contamination projects they could support. This helped the TREECS team to focus more on the functionality of TREECS and the expertise it enables when developing projects, facilitating further growth into deployment domains beyond training ranges and munitions fate and transport.

The TIM business model success for TREECS is also helping EL to consider a customer base that is more diverse than the context of the initial development of new products and services/capabilities. For TREECS, the product was mature, having been developed to monitor fate and transport at training ranges, and then was applied to other domains through the TIM process. The project sponsors noted that the TIM business model has helped them to think more broadly about the customer base when developing products and services/capabilities in the first place. As a result, EL researchers can think more dynamically during the development process and can incorporate functionality into products and services/capabilities up-front that are useful to a broader range of customer bases. This can also help these development projects to identify projects and sources of funding that are complementary to the primary development track and help the teams creating these tools to better support their development.

As a result of the successes that EL achieved in applying the TIM business model to TREECS, EL researchers and project managers are applying the TIM model developed to several new EL projects. These include several effective programs that support the soldier including life cycle analysis for existing and emergent military unique chemicals, environmental toolkit for expeditionary operations, insensitive munitions wastewater treatment, bio-inspired materials and sensors, environmental consequences of nanomaterials, and unexploded ordnance mitigation, all at varying degrees of development with distinctly different customers. In the area of water resources, the TIM business model is being applied to Infrastructure projects, threatened and endangered and invasive species, ecosystem restoration, and ecosystem goods and services. Additionally, the approach is being applied to designate technology transfer programs such as the Wetlands Regulatory Assistance Program, Dredging Operations Technical Support and Water Operations Technical Support programs.

**Acknowledgments** Special thanks to Dr. Elizabeth Fleming for her contributions to this chapter. Dr. Fleming is the director of the Environmental Laboratory at the U.S. Army Engineer Research and Development Center (ERDC) in Vicksburg, Mississippi.

## References

Bacon, F. R., & Butler, T. W. (1973). *Planned innovation: A dynamic approach for selecting, evaluating, and guiding successful development of new products*. Industrial Development Division, University of Michigan, Institute of Science and Technology.

Bacon, F. R., & Butler, T. W. (1998). Achieving planned innovation: A proven system for creating successful new products and services. Simon and Schuster.

Bozeman, B. (2000). Technology transfer and public policy: A review of research and theory. *Research Policy, 29*(4), 627–655.

Decision Partners (2011, August). Improving performance of technology transfer, technical competence and knowledge management within the USACE: Phase I. Technical report prepared for U.S. Army Engineer Research and Development Center, Pittsburgh, PA.

USACE. (2014a). Civil works. Retrieved April 10, 2014 from http://www.usace.army.mil/Missions/CivilWorks.aspx.

USACE. (2014b). Military missions. Retrieved April 10, 2014 from http://www.usace.army.mil/Missions/MilitaryMissions.aspx.

# Part III
# Applications in Other Contexts and Industries

# Chapter 7
# Farmers' Decision Making to Avoid Drug Residues in Dairy Cows: A Mental Modeling Case Study

**Sarah Thorne and Gordon Butte**

## The Opportunity

The topic of drug residues in dairy cows, like so many other issues that involve risk and risk perception, is inherently complex. Often, when addressing risk issues, the cause of the problem can be difficult to define. Similarly, it can be difficult to determine the scope and boundary of the problem.

To respond to these challenges, it is important that the project team begin with a clear, focused, and agreed upon understanding of the risk problem or opportunity to be addressed and the expected benefits to be derived from addressing it. Appling the Strategic Risk Communications™ Process begins with formulating an Opportunity Statement. The Opportunity Statement is used to:

(a) Focus the team on the full context, purpose, and value of addressing the issue or challenge at hand.
(b) Ensure the team has a shared understanding of the scope and mandate of their project.
(c) Ensure the team's efforts are achievable and measurable.

The Opportunity Statement should be outcome focused and clearly identify what is to be accomplished by when. Simple, nontechnical language should be used.

---

This chapter is adapted from Decision Partners' Final Report entitled "Farmers Decision Making to Avoid Drug Residues in Dairy Cows" (December 2011) prepared for the FDA CVM under Contract Number: HHSF223201010130A. Special thanks to Steve Ackerlund, Katherine Sousa, and Martine Hartogensis for their contributions to this chapter.

S. Thorne, M.A. (✉)
Decision Partners, 1084 Queen Street West, #32B, Mississauga, ON, Canada, L5H 4K4
e-mail: sthorne@decisionpartners.com

G. Butte
Decision Partners LLC, Suite 200, 313 East Carson Street, Pittsburgh, PA 15217, USA
e-mail: gbutte@decisionpartners.com



M.D. Wood et al., *Mental Modeling Approach*, Risk, Systems and Decisions,
DOI 10.1007/978-1-4939-6616-5_7

In considering the risk communication "opportunity" regarding drug residues in dairy cows, the CVM Project Team recognized that dairy cows sent to slaughter were disproportionately represented in terms of the number of cattle with drug residues above legal limits. "In 2008, dairy cull cows accounted for 8 % of all cattle harvested (excluding veal) at federally inspected plants but were responsible for more than 90 % of cattle residue violations. Violative residues were detected in 0.0001 % of all beef cows slaughtered but in 0.03 % of all dairy cows slaughtered. While the percentage of violative residues detected in harvested dairy cows appears to be small, it is 300 times greater than the percentage of violative residues detected in harvested beef cows." (Koeman et al. 2010) "From mid-2009 to mid-2010, roughly 95 % of Western drug residues came from the dairy industry" (Dickrell 2010).

FDA's Center for Veterinary Medicine (CVM) is responsible for ensuring safe levels of drug residues in animals slaughtered for human consumption. In 2010, the CVM Project Team decided to use mental models research to provide the insight required to develop a robust, science-informed risk communication strategy for addressing the issue. The CVM Team wanted to significantly reduce the occurrence of violative drug tissue residues in dairy cows offered for slaughter. In collaboration with the FDA Office of Planning's Risk Communications Staff, they undertook this science-based risk communications initiative to systematically address the decision making and behavior of dairy farmers. The development of this risk communications initiative was to be founded on expert knowledge; empirical research on influences on dairy farmers' decision making; and best practices from industry, academia, and government.

Key research questions established to achieve the Opportunity Statement were as follows:

- How do dairy farmers understand:
  (a) The benefits and risks of using veterinary drugs in dairy cows; and
  (b) The regulations and guidelines aimed at avoiding drug residues in dairy cows offered for slaughter?
- What influences dairy farmers' behavior to comply or not comply with these regulations and guidelines? More specifically:
  - What information do dairy farmers want and need in order to make well-informed decisions?
  - Who and what sources of information do they trust?
  - What channels and methods of information communications do they prefer?

### The Research Opportunity Was Defined as Follows

The U.S. Food and Drug Administration's Center for Veterinary Medicine (CVM), working with the FDA's Office of Planning, is sponsoring a research project to support the FDA in developing a strategic risk communications initiative that addresses the decision making and actions of dairy farmers. The goal is to significantly reduce the occurrence of violative drug tissue residues in dairy cows offered for slaughter. The initiative will build on existing knowledge and new empirical

knowledge generated from this project on the influences on dairy farmers' decision making, along with best practices from industry, academia, and government.
An important first step in this research is to elicit the knowledge and perspective of experts on various topics related to drug tissue residues in slaughtered dairy cattle, including the influences on dairy farmers' decisions to avoid violative drug tissue residues. We will integrate this input into an Expert Model, a graphic representation of expert understanding of the complex influences on dairy farmers' decisions and actions. The Expert Model will serve as the analytical framework for the design and analysis of subsequent empirical research with a sample of dairy farmers.

The insight gained from this research was used to develop risk communications strategies and messages for dairy farmers and other key stakeholders who support farmers, including veterinarians.

## Expert Modeling

All good research begins with a solid understanding of current knowledge and theory on the topic to be researched, and Mental Modeling no different. However, applied research to address problems involving risk and risk perception typically involves a large number of not well understood potential influences on decision making and behavior. To effectively work with such complexity, Mental Modeling uses decision models to bring discipline, structure, and focus on current understandings of the problem.

Accordingly, this research on drug residues in dairy cows began with the development of an *expert model*. An expert model summarizes and integrates the knowledge of *experts* on a given topic, typically in a graphic form. Within the model, arrows are used to represent *influences* that link related *nodes* or variables in the system. An arrow between two nodes means that the node at the arrow's tail may exert some *influence* on the node at the arrow's head. A completed expert model typically involves many interrelated influences that constitute an analytical framework for understanding the problem. It also becomes the basis for the design, implementation, and structured analyses of in-depth, semistructured interviews, known as *mental models* interviews, which are conducted with a sample of individuals representing a stakeholder population of interest. The general intent of the interviews is to compare and contrast a target stakeholders' understanding of the problem with that of the subject matter "experts." Since applied risk problems are often poorly understood, the notion of experts must be appropriately considered, and the expert models generally seek to be inclusive of the sometimes diverse views of multiple experts.

For this project, Decision Partners worked closely with the CVM Team and other experts to develop a Base Expert Model of Influences on the Avoidance of Violative Drug Tissue Residues in Dairy Cattle. The Model provided a structured understanding of experts' perspectives on dairy farmers' decision making regarding the avoidance of

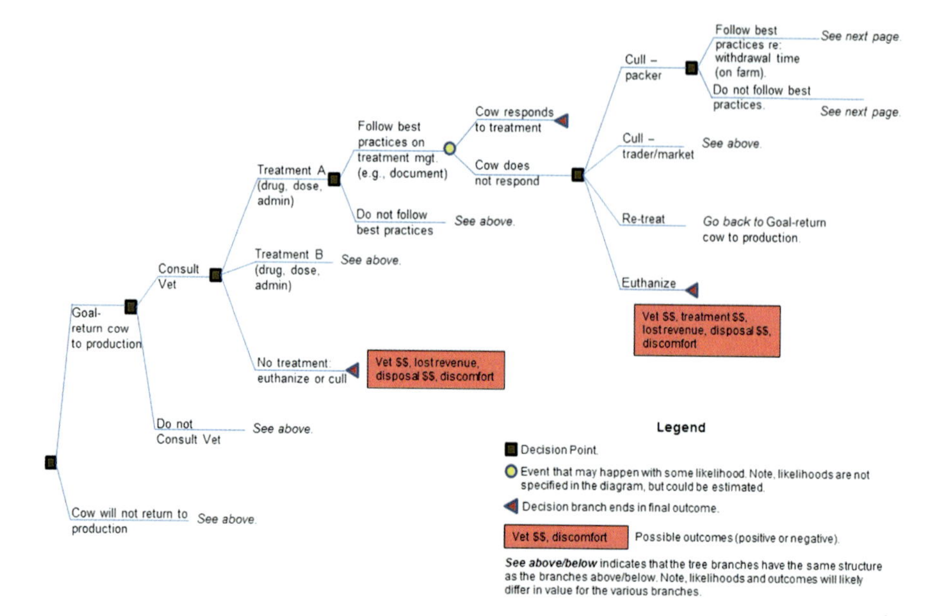


**Fig. 7.1** Decision tree of individual farmers' management decisions on the farm

violative drug tissue residues in their dairy cattle and the potential outcomes of their decision making. The model was informed by a literature review, facilitated expert workshop, expert interviews, and a validation webinar with the participating experts. Approximately 20 experts participated in the workshop, research interviews, and/or webinar including officials from FDA, USDA, and state regulatory agencies, as well as veterinary practitioners, trade professionals, meat processors and buyers, dairy operators, and academic researchers. The Model provided the analytical framework for the design, implementation, and structured analyses of the in-depth, semistructured mental models interviews with dairy farmers.

Four models were developed in the expert modeling phase of the research. Two decision tree models identify the sequence of decisions that farmers and meat packers make to avoid drug residues in dairy cows (Figs. 7.1 and 7.2).

As illustrated in Fig. 7.1, six decision points are made on the farm:

1. Is treatment needed to return a cow to milk production?
2. Does a veterinarian need to be consulted?
3. What treatment methodology will be used (e.g., drug, dose, administration)?
4. What practices are followed on treatment management? (e.g., record keeping, cow marking and separation, etc.)?
5. If treatment fails, should a cow be culled and sold for meat, retreated, or euthanized and disposed of as waste?
6. If culling for meat, what practices to follow (re: withholding, who to sell to, etc.)?

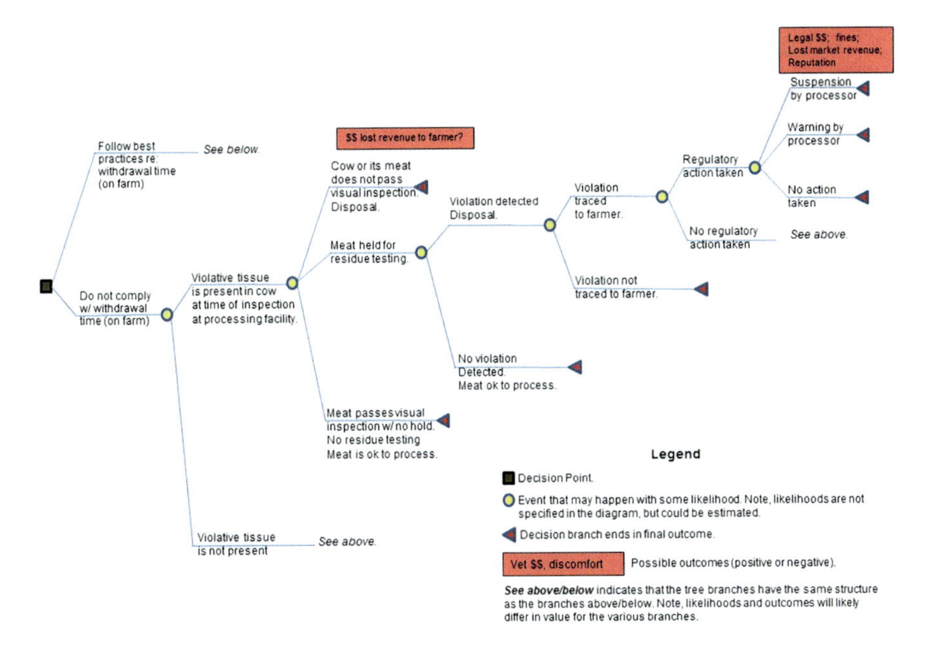


**Fig. 7.2** Decision tree of decision outcomes

An additional six decisions are involved in inspecting and responding to a violation (Fig. 7.2). These decisions occur at the meat packing plant and do not involve the farmer.

The decision trees show *what* decisions the farmer makes, but the decision trees do not tell the full story of *how* those decisions are made. Two expert mental models were developed to identify influences on the various decisions that farmers make. The *Base Expert Model of Influences on the Avoidance of Violative Drug Tissue Residues in Dairy Cattle* (Fig. 7.3) presents, at the highest level, the potential influences on those decisions, and the possible outcomes of those decisions. For example, it shows how government, industry, veterinarians, dairy farmers, traders, and meat packers all play a role in avoiding violative tissue from entering the food supply.

A *Detailed Sub-model of Influences on an Individual Farmer's Assessment of Benefits and Risks of Treatment and Management* was constructed to identify in greater detail the influences on an individual Farmer's assessment of the benefits, risks, and costs of treatment (Fig. 7.4). For example, Fig. 7.4 shows a number of potential influences, such as *Farm Practices*[1] and *Farm Culture* that can have an overall impact on the *Farmer's Capacity* to take appropriate actions in avoiding drug tissue residues. The Detailed Sub-model is aligned with the Base Model, but not all Base Model nodes appear in the Detailed Sub-models. The Detailed

[1] Common practice in describing expert models is to use first letter capitalization to indicate when referring to a specific "node" or concept in the model.

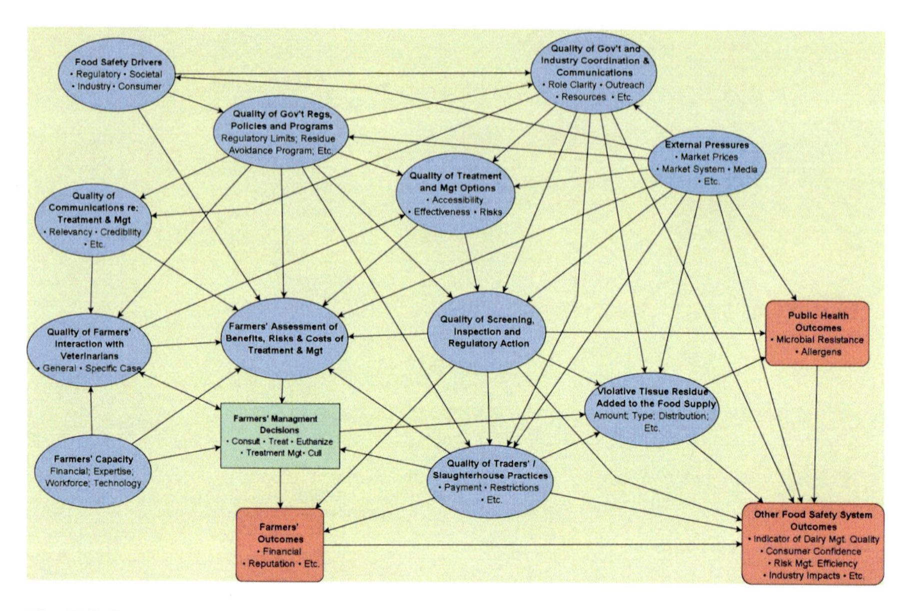


**Fig. 7.3** Base expert model of influences on the avoidance of violative drug tissue residues in dairy cattle

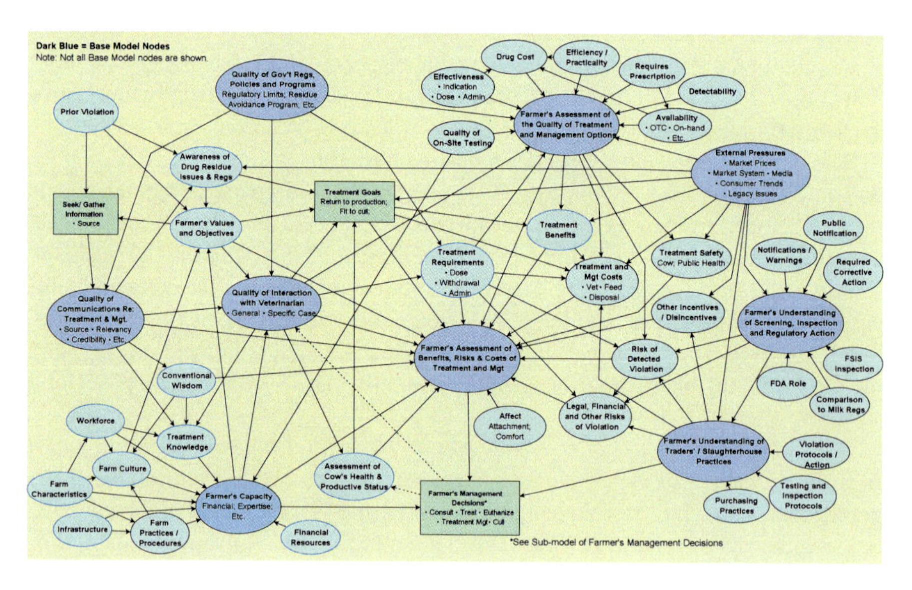


**Fig. 7.4** Detailed submodel of influences on an individual farmer's assessment of benefits and risks of treatment and management

Expert Model is also aligned with the decision tree (Figs. 7.1 and 7.2). It supplements the decision tree by showing a broader range of influences on farmers' decision making.

The process of developing and discussing the collective set of expert models generated a number of additional topics that were considered along with the original research questions in the development of the interview protocol:

- Farmers want to do the right thing, but they lack awareness and understanding of the significance of the issues because...
  - They may not view their animal as a meat product.
  - Production-driven economics may be a key driver.
  - Slaughter may be construed as treatment failure.
  - Once a cow leaves a farm, it may no longer be viewed as a farmer's responsibility.
  - "People take penicillin, what's the big deal?"
  - "It's a drop in the bucket, it doesn't matter."
- Drug treatment (and its labeling) is very complicated:
  - Dairy cattle are more prone to illness than beef cattle.
  - Veterinarians may not be consulted as often as necessary.
  - Farmers may use whatever drugs are on hand.
- There are limited viable options to culling:
  - Disposing of treated cows may be considered toxic waste.
  - Rendering facilities won't take cows more than 5 years old.
- There are significant capacity issues relating to economic, record keeping, and worker training constraints. Moreover:
  - Treatment management practices are not standardized across the industry.
  - Decisions can be made by one person and carried out by another.
  - They have learned bad habits and are guided by outdated conventional wisdom.
- "Bad actors" generally are poor managers, have other problems, and are "living recklessly."

## Mental Modeling

Mental models interviews were conducted with a sample of dairy farmers. In this case, the research focused on two cohorts:

- Non-Violators: individuals were sought representing dairy operations that are either known to not have experienced a violation or whose compliance status is unknown.
- Violators: individuals were sought representing dairy operations that are known to have experienced a regulatory violation. The goal was to recruit those who had a violation within the past 3 years.

## Sample Development

To develop an appropriate sample, individuals who serve as a key decision-maker regarding treatment and/or culling at the dairy operation, such as dairy herdsman, production manager, or safety manager, were sought. For small operations, this included the dairy owner. FDA used internal databases on Violators and reached out to its state government partners to identify Non-Violators. The FDA Team provided Decision Partners with a sample of 370 Dairy Farmers.

## Protocol Design

A semistructured interview protocol was developed in consultation with the CVM Team, designed to gain the insight required to address the Team's research questions. The protocol included questions designed to solicit the Dairy Farmer's thinking about key topics identified in the expert model. Interviewees were offered an agenda of topics to discuss in a way that allowed for free expression. Interviewees were also encouraged to raise additional topics spontaneously and to elaborate on their perspectives. The interview protocol was organized as follows:

- Part 1: Farming operations and the Farmer's (Interviewee's) role on the farm.
- Part 2a: How the Farmer or his or her staff make decisions regarding the medical treatment or culling of their dairy herd.
- Part 2b: The Farmer's thinking about the topic of avoiding violative drug tissue residues in dairy cows sold into the meat supply for beef.
- Part 3: The Farmer's thoughts on the types and sources of information that can help Dairy Farmers when treating and culling dairy cattle.

## Sampling Process

In-depth phone interviews were conducted with 15 Violators and 15 Non-Violators between April 2, 2011 and November 21, 2011. These included 9 initial pretests, following which small protocol refinements were made to improve delivery. Interviews ranged from 29 to 68 min, averaging 50 min. Information obtained through interviews was held as confidential. No identifying information about the Interviewees or their operations was provided to the FDA, and no comments are attributed to any Interviewee in this Report. Interviews were recorded with the Interviewee's permission. A $50 honorarium was awarded to interviewees.

Interviews were conducted with Dairy Farmers in numerous states across the country. Eleven of the fifteen Interviewees who had violations indicated that they occurred in the past 3 years. Table 7.1 provides further detail on the violations represented in the sample.

Among those interviewed, farm sizes varied from family-owned operations involving 45 milking cows to larger scale businesses involving 6500 cows. Some interviewees described other aspects of their operation including foraging acreage

**Table 7.1** Characterization of violations

| Violator | Herd size[a] | Number of violations | Years since violation | Drug |
|---|---|---|---|---|
| 1 | Medium | -- | <3 | -- |
| 2 | Small | 1 | <3 | Penicillin |
| 3 | Small | -- | >3 | -- |
| 4 | Small | -- | >3 | -- |
| 5 | Medium | 1 | <3 | Penicillin |
| 6 | Small | 1 | <3 | Sulfamethazine |
| 7 | Very large | 1[b] | <3 | Sulfadimethoxine |
| 8 | Medium | 1 | <3 | Desfuroylceftiofur |
| 9 | Very large | -- | <3 | -- |
| 10 | Large | 2 | <3 | Desfuroylceftiofur<br>Desfuroylceftiofur |
| 11 | Very large | 10 | <3 | Gentamicin<br>Penicillin<br>Flunixin<br>Penicillin<br>Penicillin<br>Penicillin<br>Penicillin<br>Oxytetracycline<br>Penicillin<br>Sulfadimethoxine |
| 12 | Medium | 1 | <3 | Flunixin<br>Penicillin |
| 13 | Large | 2 | <3 | Tetracycline<br>Penicillin |
| 14 | Medium | 1 | >3 | Gentocin |
| 15 | Large | -- | >3 | -- |

(--) dash indicates Violators identified by chance during interviews and data on number of violations and which drugs were not systematically identified

[a]Small (<100 milking cows), medium, (101–500 milking cows), large (501–2500 milking cows), very large (2500 milking cows)

[b]Self-reported additional violations during the interview and stated that FDA records are not accurate

of crop land, "custom chopping." One farmer described his/her farm as a "diversified agricultural enterprise," which in addition to having dairy and other cattle, produces and sells maple syrup products and runs a bed and breakfast. One operation was an organic dairy farm and another described his/her farm as "holistic."

## Coding and Analysis

Interviews were transcribed and then coded against the Expert Model by trained coders who followed a standard set of coding and analysis procedures. Coded interviews were transferred to a database where detailed analyses were performed to identify detailed themes and their relationships. Structured qualitative analysis of the interviews against the Expert Model was used to identify the key areas of alignment and critical gaps in the thinking of experts and the farmers.

## Key Results

The following is a top-line summary of key findings from the mental models research conducted with 30 dairy farmers from April to November 2011.

All of the interviewees emphasized that protecting the food supply by avoiding drug residue violations was very important: "Dairy farmers want to do what is right. There are a handful of bad apples that spoil it for all of us." This research demonstrated that Dairy Farmers share a critically important common interest with FDA—sustaining public trust in a high quality and safe food system. They clearly recognized their role in keeping their products—both milk and meat products from dairy cows, safe:

> America has always had the safest and regulated food production, and people respect that and so we want to maintain that. Nobody is going to want to buy your product if it is poor quality or contains potential harmful antibiotics.

As shown in Fig. 7.5, almost all said they were very or somewhat familiar with the rules and regulations related to drug tissue residues in dairy cattle intended for the food stream. All interviewees, both Violators and Non-Violators, believed the rules and regulations were very or somewhat appropriate (Fig. 7.6) and commented that they are doing their best to adhere to them with the guidance and support of their veterinarians. Most also said that they have the information they need to make good decisions and nearly all stated that they are following specific cow treatment

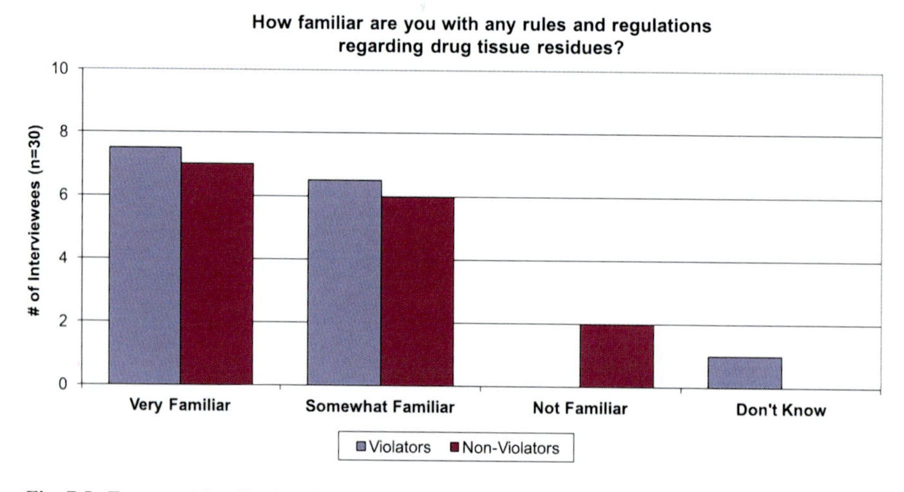


**Fig. 7.5** Expressed familiarity with rules and regulations. The fractions represent one Interviewee who said "between Very and Somewhat Familiar"

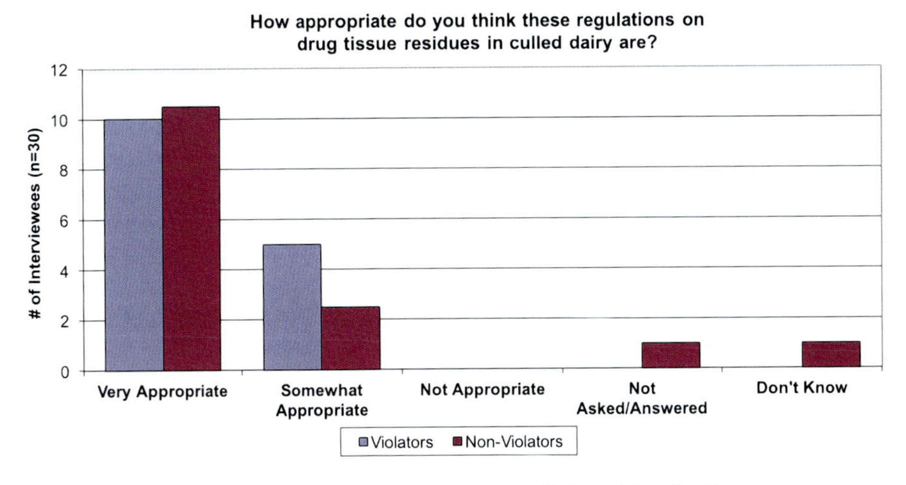


**Fig. 7.6** Expressed appropriateness of rules and regulations. The fractions represent one Interviewee who said "between Very and Somewhat Appropriate"

and management procedures they have put in place in their operations to avoid drug residues "very well."

To care for sick cows, the interviewees noted that veterinarians are trusted and respected partners who work closely with them to write protocols specific for their farm detailing appropriate treatment options, specific dosages, administration, and withholding periods. Veterinarians typically visit the farms regularly, are available in severe situations, and/or are on call to provide treatment advice or answer questions. For most, working with veterinarians is an important part of the operations.

Dairy Farmers who had developed written protocols for managing cow treatment said they routinely follow them. The protocols limit the number of drugs to reduce complexity and maximize those with shorter withholding times. Methods for documenting drugs used and for managing treated cows are also defined in these protocols. Many described having either a written protocol that defines cow treatment practices on their farm, or routinely using forms and procedures to document cow treatments. Some mentioned they did not have formalized standard operating procedures, noting, "we just kind of do what we think we've got to do." Importantly, a range of treatment and cow management practices were described, suggesting there is little common knowledge on what constitutes "best practices" within the industry.

Nearly all Dairy Farmers reported following label instructions pertaining to dosage, administration, and withholding time. Many Interviewees mentioned following "standard protocols" when selecting drugs, dosages, and routes of drug administration. Some mentioned that they consistently follow label instructions. However, some mentioned not following a standard protocol for one or more parts of their treatment: "I can think of times where maybe a vet might have said 'okay, I want you to really shock this cow with more than what the label says.'" A few, including Violators and Non-Violators, mentioned that they use some drugs off-label, particularly penicillin; others mentioned that they now avoid this practice. Many Non-Violators and a few Violators said they may hold a cow longer than the designated withholding time to ensure that no residues are left in the cow, particularly for a very ill cow that is suspected of having a slower drug metabolism. A few Interviewees emphasized prevention to avoid treatment, including vaccinations, more frequent cleaning, and making changes to the barn/stall infrastructure.

Dairy Farmers reported that on average, they cull 10–40 % percent of their herds annually. Generally, culling because of nonresponsiveness to drug treatment is a relatively infrequent occurrence. Rather, cows are culled because of age, overall health, production rates, and reproduction ability. Therefore, culling because of drug residues is only one consideration, and not a consideration that is as frequently applicable as others.

The research revealed that Dairy Farmers were generally unaware of trader/slaughterhouse practices, how violations are detected, and the penalties for violations; however, they do not believe they are constrained by this lack of knowledge when making treatment or culling decisions. Whether they thought detection was likely or not, a common theme among Interviewees from both cohorts was that they did not want to have a violation: "I have to think that every cow would be detected. I have to think that. It will make me more careful on my end."

The explanations of treatment practices generally underscored Dairy Farmers' focused efforts to avoid sending dairy cows with residual pharmaceuticals to slaughterhouses. However, Dairy Farmers perceived that complete elimination of all drug residue violations is not possible because of variability among cows in the rate of drug elimination (particularly for sick cows), the potential for infrequent mistakes, and the lack of a test method that can be used on the farm to ensure compliance. While Farmers think complete elimination of violations is not possible, they are supportive of actions against repeat violators: "If you got somebody that is a repeat violator, the dairy industry doesn't want them around either."

A key finding of this research is that there were no substantial differences in practices or decision-making behavior of Violators and Non-Violators. The differences noted between those who had experienced a violation and those who had not were few in comparison to the similarities with generally only modest differences in frequency of mention toward a particular side of an issue. As shown in Fig. 7.7, Violators were more likely to believe that it is "very likely" that culled cows containing residues would be detected during inspection; were more likely to mention having a comprehensive, written farm protocol; were less likely to have or use the Internet, and more frequently stated that they consult labels "very often." Non-Violators were more likely to mention that they hold cows longer than the prescribed withholding times. Some with past violations did mention that they have since improved their treatment management methods. However, nearly all talked about having improved their practices compared to what they have been doing in the past in response to increased consumer awareness and concern for residues.

Among those interviewed, the occurrence of violations was more common for larger farms. Given the size of their herds and their years of operations, some with more than one violation characterized their violations as infrequent. One Violator,

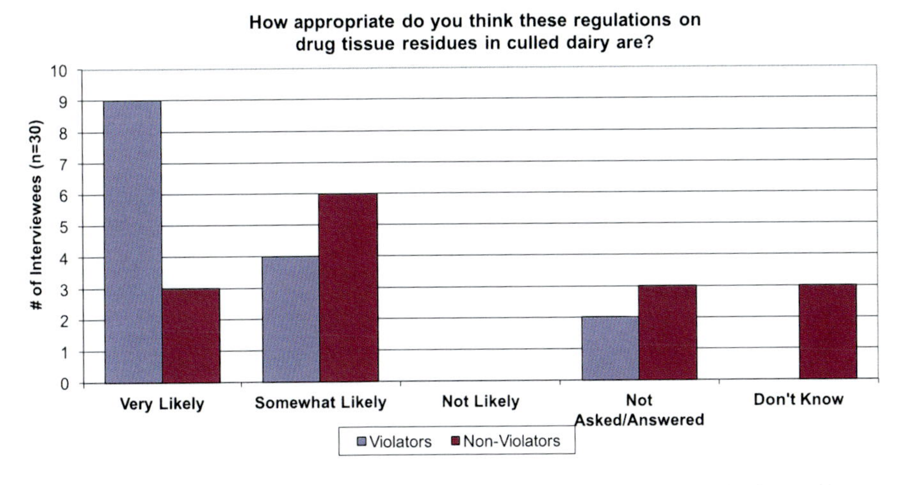


**Fig. 7.7** Farmers' assessment of likelihood of residue detection in culled cows having residues

who has 600 dairy cattle, noted that he has had only two violations in over 30 years. Another, whose farm had been in the family for over 100 years, commented that they have 100 cattle and have had one violation 6 or 7 years ago.

Despite their best efforts to avoid drug tissue residues, some commented that "accidents do happen" and there will always be a risk—"an element of chance"—that drug tissue residues may remain: "Even if you have the strictest protocols in the world, it is still only 99.9 % safe."

While nearly all Dairy Farmers said "they want to do what is right," challenges remain. When asked to comment on the most likely source of violations, reasons offered in descending order of frequency of mention were as follows:

- Lack of treatment documentation.
- Off-label drug use.
- Human error in treatment documentation or communication.
- Questionable motives, i.e., "a few bad apples."

Additionally, Interviewees' statements suggest additional reasons for violations:

- A few may not be motivated to achieve standards because they do not understand or believe in the basis for the standards.
- A few stated that they do not trust or find adequate value in working with their veterinarian.
- Lack of "industry standards" or "best practices" to avoid violations.
- Uncertainty on withholding times for sick cows:

> There was one about a year ago that I treated for pneumonia and she didn't respond or she responded partially but it didn't clear completely. In that case, I called the vet and he said it's a growing problem that won't go away, so I kept the animal according to the label

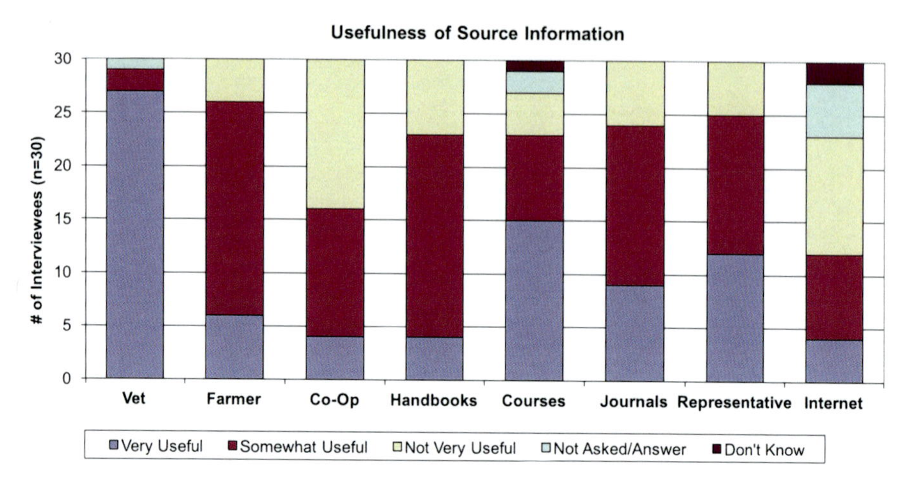


**Fig. 7.8** Farmers' rating of information sources

> required withhold time before I culled her, but I did get caught that time. I guess, probably because of the condition of the animal she wasn't healthy enough to pass off antibiotics in the normal time period.

Farmers rated their veterinarian as the most useful source of information, particularly for new information and technically complex topics (Fig. 7.8). Other sources of information, such as other drug representatives, other Farmers and the Internet appear to be of secondary or limited importance and/or trustworthiness.

Dairy Farmers use a wide range of information sources such as journals, other farmers, drug representatives, and the Internet; however, they rely most frequently upon two sources:

- They tend to trust and rely on their veterinarians for new information and technically complex topics, and
- They rely on drug labels for routine checks on procedure, including withholding times.

Dairy farming is a competitive, challenging business and farmers have limited time for nonfarming tasks. Generally expressed communication preferences are as follows:

- Face-to-face, two-way dialogue to solve problems and address new and important topics. Veterinarian involvement desirable.
- Bulletins, fact sheets, etc., for routine updates (e.g., violation rates) or to explain well accepted facts (e.g., regulatory requirements).
- Courses, if focused on topics of interest and convenient to accommodate farmers' work schedules.

In closing, Farmers were asked to offer one piece of advice regarding how FDA could help farmers avoid residue violations. In general order of frequency of mention, the advice was:

- Engage in two-way dialogue.
- Provide education.
- Develop and communicate clear guidelines on "best practices."
- Recognize accidents and be tough on repeat violators.
- Develop a meat residue test to be used on the farm.
- Promote prevention.
- Improve culled cow tracking procedures with traders and slaughterhouses.
- Keep up the good work: "They're doing the right thing...and we should get our act together."

## Improving Risk Communication

This research supports FDA's risk communications initiative by providing insights gained from in-depth interviews with Dairy Farmers that can be used to develop focused risk communications strategies and messages for dairy farmers, and other key stakeholders, such as veterinarians. Table 7.2 provides a disciplined approach for distilling the key insights that are most pertinent to developing an informed risk communication strategy. A list of guiding questions supporting strategic risk communications provides insight on what Farmers know, don't know, want to know, and what communications processes they trust.

## Variances Between Violators and Non-violators

A key finding of this research is that there were no substantial differences in practices or decision-making behavior of Violators and Non-Violators. The differences noted between those who had experienced a violation and those who had not were few in comparison to the similarities with generally only modest differences in frequency of mention toward a particular side of an issue. The most substantial differences expressed include:

- Violators were more likely to believe that it is "very likely" that culled cows containing residues would be detected during inspection.
- Violators were twice as likely to mention that the protocols used on their farm were developed with the veterinarian.
- Violators more frequently stated that they consult labels "very often."
- Some Violators commented that they either "have no Internet access" and/or are not computer literate.
- Many Non-Violators and a few Violators described withholding their cows longer than the stipulated withholding times.

One other variance noted was that Farmers from small farms tended to describe less formal systems for documenting and managing cow treatments. Only two of the

**Table 7.2** Mental models assessment summary

| Guiding questions | Dairy farmers: violators and non-violators |
|---|---|
| What do they value? | Dairy farmers are focused on providing safe, quality food products that are trusted by consumers. They value their role in the food system, including their:<br>• Rep3utation—their own and that of the industry<br>• Responsibility to public health and safety: "Most farmers really want to do what is right" |
| What do they know now about the topic/risk that is correct? | Dairy Farmers expressed detailed knowledge of:<br>• When treatment is needed and when to consult their veterinarian<br>• Selection of drugs that minimize or eliminate withholding times<br>• Importance of adhering to methods for drug administration and withholding times listed on drug labels<br>• Importance of observing withholding times<br>• Importance of keeping detailed records for each cow treated<br>(Their behavior is consistent with the 6 Decision Steps pertaining to decision made on the farm, illustrated the decision tree model, Fig. 7.1) |
| What are the key factors that they consider when making culling decisions? | Culling because a cow does not respond to treatment is less frequent than culling for other reasons. Key factors considered in deciding to treat or cull are:<br>• Cow health, productivity, age, ability to breed<br>• Economics—input costs vs. productivity<br>• Withholding times for treated cows<br>• The need for caution—uncertainty in residue levels, particularly for very sick cows that might metabolize drugs more slowly, and "Accidents do happen"<br>• Knowledge that drug tissue residues are important to customers<br>• Commitment to and concern about public health, industry reputation, and their business success<br>• FDA rules, which are regarded as necessary and generally appropriate |
| What don't they know or misunderstand that is consequential? | Dairy Farmers expressed limited knowledge of decisions made by others once a cow is culled, including the consequences of a violation; however, this knowledge does not appear to be consequential to treatment and culling decisions. Key questions expressed by Farmers as consequential to their decisions include:<br>• Are withholding times applicable for sick cows suspected of slower drug metabolism rates?<br>• How are meat residue standards derived? Clarify the health protection goals used to establish standards. Reconcile interest in safe, untainted food and the technical basis for an allowable residue |

(continued)

**Table 7.2** (continued)

| Guiding questions | Dairy farmers: violators and non-violators |
|---|---|
| What do they want/want to know? | Beyond specific technical questions that are consequential to their decisions, Dairy Farmers wanted to know:<br>• Best practices—they suggested FDA develop and communicate those practices that are known to produce results<br>• Rules and regulations—beyond drug administration and withholding time requirements, they suggested FDA could make important points on rules and regulations more accessible and understandable. For example, rules pertaining to feeding milk from treated cows to calves<br>• On the farm tests—they suggested FDA could develop a test to determine if cow tissues have residues, as is available for milk<br>• If FDA listens to and understands dairy farming operations and challenges |
| Who do they trust and why? | Dairy Farmers use a range of information sources such as journals, other farmers, drug representatives, and the Internet; however, they rely most frequently upon two sources:<br>• They tend to trust and rely on their veterinarians for new information and technically complex topics, and<br>• They rely on drug labels for routine checks on procedure, including withholding times |
| What communication methods do they prefer? | Farmers noted that dairy farming is a competitive, challenging business and farmers have limited time for nonfarming tasks. They prefer:<br>• Face-to-face, two-way dialogue to solve problems and address new and important topics. They prefer their veterinarians to be involved in such discussions<br>• Bulletins, fact sheets, etc. for routine updates (i.e., on violation rates) or to explain well-accepted facts (e.g., regulatory requirements)<br>• Courses, if focused on topics of interest and convenient to accommodate farmers' work schedules |
| What do they recommend—their advice to FDA? | Farmers support FDA's efforts to ensure public confidence in a safe food supply, but they want to be treated respectfully regarding infrequent violations and seek meaningful involvement in future development and communication of "best" practices for minimizing drug residue violations. Citing lack of documentation, off-label drug use, human error, and questionable motives as principle causes of violations, Dairy Farmers suggested the following solutions:<br>• Engage in two-way dialogue<br>• Provide education<br>• Develop and communicate clear guidelines<br>• Recognize accidents happen, but be tough on repeat violators<br>• Develop a drug residue test (for meat) to be used on the farm<br>• Promote prevention<br>• Improve culled cow tracking procedures<br>• Keep up the good work |

six Farmers on small farms (less than 100 milking cows) reported using written procedures for documenting cow treatment and none of these mentioned having developed a standard written protocol for cow treatment.

## Considerations on Next Steps for Strategic Risk Communications with Dairy Farmers

The findings of this research demonstrate that nearly all Dairy Farmers share a critically important common goal with FDA—sustaining trust and confidence in the quality and safety of America's food system. The research indicates that a strong majority of Dairy Farmers are committed to their role in providing safe products from their operations. They understand the benefits and risks of using veterinary drugs in dairy cows, are familiar with key aspects of regulations pertaining to drug usage and withholding times, and are motivated to do what is necessary to achieve compliance. The clear prevalence of shared values and goals between FDA and Dairy Farmers on food safety suggests an opportunity for a collaboration-based risk communication strategy.

When the Decision Partners team presented the research to the CVM Team and interested colleagues in December 2011, they offered the numerous considerations. They suggested the CVM Team build on the results of this research by engaging dairy farmers, veterinarians, and other key industry stakeholders in a process to define guidance on best practices that are achievable on the farm and can be demonstrated to further reduce the possibility for unintended violation of drug residue standards for foods. This collaborative effort could build on the collaboration undertaken in the research phase of this initiative, namely, the engagement of expert stakeholders in the development of the expert model, and the outreach to dairy farmers through the process of conducting mental models research.

Drawing on strategic risk communications principles and practices the development of such guidance could:

- Demonstrate FDA's commitment to working with dairy stakeholders to ensure safe levels of drug residues in animals intended for human consumption.
- Clarify and communicate FDA food safety goals and how the goals relate to the need for and scientific basis of drug residue standards in animals intended for human consumption.
- Enable participating stakeholders to come to a shared understanding of best management practices for avoiding drug residues at all stages of the meat production process.
- Enable investigation into possible on-site testing methods that would enable dairy farmers to test for residual drugs before they release cows for sale to traders and slaughterhouses.
- Assess possible on-site testing methods that would enable dairy farmers to test for residual drugs before they release cows for sale to traders and slaughterhouses.
- Identify and address risk in the farm to slaughterhouse chain of custody.

- Support the development of clear messages and materials, that once pretested with focal stakeholders, could be communicated broadly to the dairy industry through FDA, participating stakeholders, and their organizations, veterinarians, University Extension, industry associations, and a variety of other channels including FDA's website.
- Support the development of other outreach initiatives, including training, which could be conducted by FDA, possibly with other stakeholders.
- Update the expert models to reflect decision making and behavior in a complex system.
- Revise strategy and communications content associated with violations.

## Key Learnings and Applying the Results

Following a presentation of the research results to the CVM Team and many of their colleagues in December, 2011, the Team used the findings to address specific challenges identified in the research and to develop focused communications strategies.

When asked about the key learnings for the CVM Team, project lead Martine Hartogensis commented that "the project really opened our eyes to things we were missing and things we could be more clear on in our communications. For example, we had to reframe our thinking on residues and residue avoidance. And we had to develop specific communications strategies that met the needs of our various stakeholders on the issues and challenges identified. Using the research results to open the dialogue is a new approach for CVM and it is really having a significant impact."

From a risk management perspective, the CVM Team reached out to the organizations responsible for the marketing and sales of the various penicillin drugs used by dairy farmers and shared their concerns about the adequacy of the labels. Some of these drugs have been in use since the 1950s. Many are sold over the counter. Based on the research results, the Team asked these organizations to meet to discuss approaches for working collaboratively to address the concerns.

A significant change for CVM was the strategy for conducting outreach to drug sponsors. The Team decided to use a dialogue approach rather than "the traditional us telling them what we want them to do," noted Hartogensis. The Team reached out to the sponsors in small group meetings or by phone to ask for their help, voluntarily. They framed the challenge of residual pharmaceuticals in dairy cows as a public health issue. To date, Hartogensis reports that sponsors have been very positive and supportive of achieving better public health outcomes through more and better communication about the proper use of these fundamental drugs. "They clearly saw the need and are willing to work with us on how to better communicate with users. And they appreciated this new, more collaborative approach."

The CVM Team has also changed its strategy for working with veterinarians, reaching out to them as partners. Both formal and informal communications channels have been used to share the results of the research and the key learnings and ask veterinarians to support CVM's efforts in improving public health outcomes through more effective outreach and communications with producers. Hartogensis commented: "We've emphasized one of our key learnings—that producers really want to do the right thing, but they need the right tools. We're asking vets to help by partnering with producers and provide them with better communications aligned to their needs and at a level they can understand." To date, the response has been very positive. Hartogensis believes this is largely due to the CVM's new approach of outreach and partnership, underscored by sharing the critical insight revealed by the research with vets who are a primary point of contact with producers and highly trusted.

The CVM Team will continue to share research results throughout CVM and with various stakeholder groups through presentations, meetings, and peer-reviewed papers. One group, CVM and external veterinarians working with swine producers, is using the results to frame the dialogue on how the key stakeholders can work together to reduce residual drugs in swine.

Summing up the key learnings from applying the Mental Modeling approach to the challenge of residual drugs in dairy cows intended for the food stream, Hartogensis said: "We all learned a lot from this work. Based on the final report, we are willing to give producers the benefit of the doubt, even when they have had violations in the past, as long as they take the necessary steps to be in compliance. We are all working on the same challenges, but from different perspectives. We need to think about the overall system and what needs to be done at every step in the chain of custody to achieve the public health outcomes we all desire."

## References

Dickrell, J. (2010). Dairy to blame for beef drug residues. Retrieved January 11, 2016, from http://www.agweb.com/article/dairy_to_blame_for_beef_drug_residues2/.

Koeman, J., Goldsmith, T., & Eustice, R. (2010). Now more than ever: Check records and talk to your veterinarian before you ship the cow. Retrieved January 11, 2016, from http://www.progressivedairy.com/topics/herd-health/now-more-than-ever-check-records-and-talk-to-your-veterinarian-before-you-ship-the-cow.

# Chapter 8
# Influence of the CHEMM Tool on Planning, Preparedness, and Emergency Response to Hazardous Chemical Exposures: A Customized Strategic Communications Process Based on Mental Modeling

**Daniel Kovacs, Sarah Thorne, and Gordon Butte**

## The Opportunity

The National Library of Medicine maintains the world's largest biomedical library from which the Specialized Information Services Division produces information on a range of topics, including disaster/emergency preparedness and response.[1] CHEMM is a software-supported decision tool, produced by the SIS that provides first responders, first receivers, and other potential users rapid access to comprehensive information needed to better make critical decisions when planning for, preparing to respond to, and responding to hazardous chemical exposure emergencies. The primary goal of CHEMM is to save lives through

---

This chapter is adapted from Decision Partners' Final Report entitled "Influences of the CHEMM tool on Planning, Preparedness and Emergency Response to Hazardous Chemical Exposures" (February 2012) Prepared for NLM under Contract Number: HHSN276200900787P/0002. The authors would like to thank the NLM and the other members of the NLM Project Team including Florence Chang and Jennifer Pakiam for funding support and significant expert contribution to this work. The authors also with to thank our contributor, Pertti Hakkinen, Acting Head, Specialized Information Services, Office of Clinical Toxicology at NLM, for his assistance with this chapter.

[1] http://www.nlm.nih.gov/about/index.html; http://www.nlm.nih.gov/pubs/factsheets/sis.html.

D. Kovacs, Ph.D. (✉)
Decision Partners, 1458 Jersey Street, Lake Milton, OH 44429, USA
e-mail: dkovacs@decisionparters.com

S. Thorne, M.A.
Decision Partners, 1084 Queen Street West, #32B, Mississauga, ON, Canada L5H 4K4
e-mail: sthorne@decisionpartners.com

G. Butte
Decision Partners LLC, Suite 200, 313 East Carson Street, Pittsburgh, PA 15217, USA
e-mail: gbutte@decisionpartners.com



M.D. Wood et al., *Mental Modeling Approach*, Risk, Systems and Decisions,
DOI 10.1007/978-1-4939-6616-5_8

improved response to such emergencies. CHEMM is uniquely suited to addressing emergencies with the potential for mass casualties resulting from large-scale release of hazardous chemicals used in large quantities in industrial applications, or from release of the most highly toxic substances such as those that may be used in terrorist attacks. Such scenarios are extremely unusual and typical responders and receivers are likely to have limited training and little to no experience in responding to such incidents.

In 2010, the CHEMM Team asked Decision Partners to assist them in the development of CHEMM by employing a multiphase Mental Modeling approach to gain insight into: (a) how first responders, first receivers, and other potential users think through and make decisions in a chemical exposure-related mass casualty event and (b) the role CHEMM could play in supporting these activities.

The first step in the Mental Modeling approach was to work with the CHEMM Team to develop a clear, measureable opportunity statement. To guide its work, the Team developed the following Opportunity Statement:

> *The opportunity is to optimize effective decision making in a chemical hazards emergency by enhancing user interface with the CHEMM Tool. We will do this by using insight into how first responders, first receivers and other users think through and make decisions in a chemical exposure-related mass casualty event to design and validate the CHEMM Tool.*

The multiphase Mental Modeling approach was then undertaken comprising the following steps:

- A draft expert model was developed working with a small group of NLM experts and through a review of other emergency management resources currently available.
- Detailed expert models were developed through a workshop with 27 expert stakeholders, including first responders, first receivers, and government and academic researchers. These detailed models served as the analytical framework for the mental models research.
- Mental models interviews were conducted with 40 potential CHEMM stakeholders, including first responders, first receivers, and others, including emergency response trainers and planners.
- An Update and Strategy Development Workshop was held with key stakeholders to update them on the progress of the CHEMM research and to develop strategies for future CHEMM outreach and development.
- A Stakeholder Outreach Plan was developed to continue and enhance stakeholder engagement for ongoing development of CHEMM.

The remainder of this chapter describes the research approach and presents the results of the research efforts. It concludes with a discussion of the value of this approach in Client Perspectives on Mental Models Research, Key Learnings, and Applying the Results.

## Mental Modeling Approach

Decades of research demonstrate that people's judgments about complex issues are guided by their "mental models." Mental models are the tacit webs of beliefs that all people draw upon to interpret and make inferences about issues that come to their attention. They cannot be determined without empirical research.

The Mental Modeling approach has been used effectively to understand risk management contexts and develop robust solutions that incorporate the beliefs and priorities of stakeholder groups. It is ideally suited to issues where:

- Disparate viewpoints must be synthesized on complex topics;
- Decisions are required among multiple potential risk management options with a significant degree of consequence; and
- Transparency is required when characterizing the issue, incorporating stakeholder input, and designing appropriate risk management solutions.

### *Expert Models*

Following the development of the Opportunity Statement, a base expert model, developed through review of relevant literature and through formal and informal discussions with the Project Team and a limited number of key experts, is developed. The draft base expert model is then shared with a larger group of experts and stakeholders in a more formal, moderated workshop setting where participants are encouraged to discuss important elements of the model and provide input into development of *detailed* expert models. The detailed expert models, created subsequent to the workshop, are then presented back to workshop participants, in a workshop or by webinar, to validate that the system being depicted is accurate and adequately reflects the input of participants. The validated expert models then become the analytical basis for design and analysis of subsequent mental models research with a larger group of expert or nonexpert stakeholders.

### *Mental Models Research*

Mental models research comprises one-on-one, in-depth interviews, usually conducted by telephone, in which participants respond to a series of open-ended questions, allowing them to express themselves in their own terms, thereby allowing unexpected beliefs to emerge and underlying thought processes to be revealed. Mental models interview transcripts are coded and analyzed against the expert models to ensure that all data is be accounted for. The analysis of the results yields valuable, in-depth insight into stakeholder thinking about the topic at hand.

## Expert Models of Influences on CHEMM Effectiveness

The CHEMM Expert Models were created to identify and characterize the primary influences on the effectiveness of CHEMM, supporting planning and prioritization of stakeholder outreach, and ongoing CHEMM development. The Models were also used as the analytical basis for the research conducted in this project. Draft versions of these Models were created through consultation with the CHEMM Project Team and were presented to 27 CHEMM stakeholders in a workshop conducted February 17, 2010. The facilitated workshop was designed to enable open, frank discussions on the issues and challenges that matter most to participants. The purpose of the workshop was to elicit and focus knowledge and insights in a way that would help shape the design of CHEMM.

Workshop participants represented a variety of disciplines associated with emergency response including: fire fighters, hazmat responders, physicians, federal and state agencies, industry, consulting firms, and researchers. Feedback from workshop participants was used to develop the CHEMM Detailed Expert Models, presented later, as well as a CHEMM User Matrix, detailing information needs for specific CHEMM Users.

The updated CHEMM Expert Models were presented to workshop participants in a follow-up validation web conference on May 18, 2010 and were used as the analytical basis for mental models research (presented later). They will continue to provide a framework for future research and development of CHEMM.

## Expert Model Narrative

The Expert Model Narrative guides the reader through the CHEMM Expert Models, providing descriptions of the key variables (termed *nodes*) and the relationships among the variables (termed *influences*). Two Models are presented. The first Model depicts a system perspective, a more general model providing the context within which CHEMM will be employed. The second Model depicts a user perspective illustrating: the influences that drive individuals to use CHEMM, how they use it, their desired outcomes, and the qualities that will be used to gauge the effectiveness of CHEMM.

### *Influence of the CHEMM Tool on Planning, Preparedness, and Emergency Response to Hazardous Chemical Exposures: System Perspective*

The System Perspective CHEMM Expert Model is presented in Fig. 8.1.

The Model starts in the upper left corner with Drivers (Fig. 8.2) or initiating influences on the system including influences overall needs and activities for *Public*

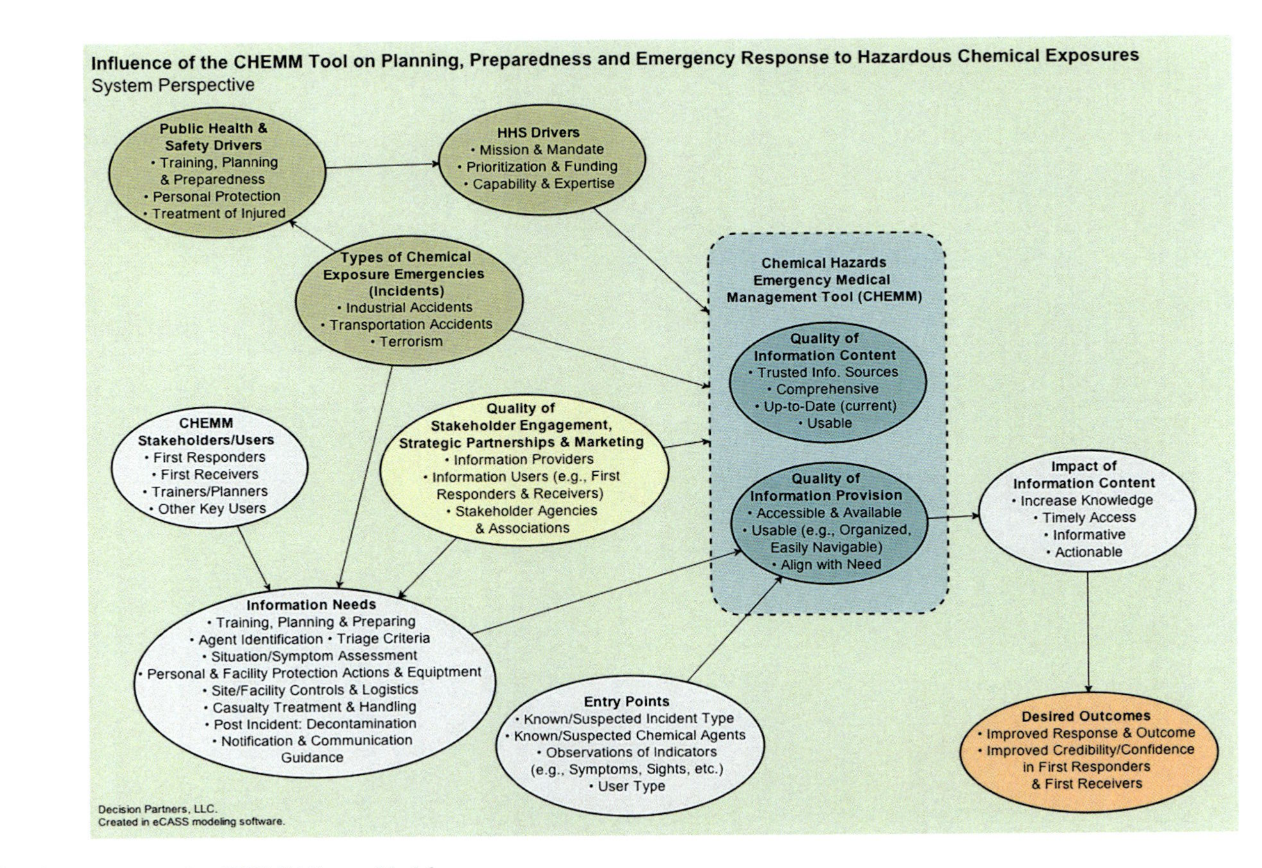


**Fig. 8.1** System perspective CHEMM Expert Model

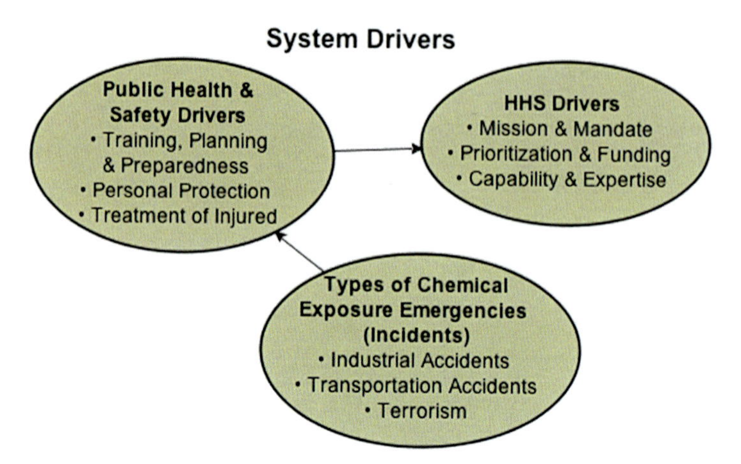


**Fig. 8.2** System Drivers

*Health and Safety*, influences from the *Department of Health and Human Services*, and influences related to the *Types of Chemical Exposure Emergencies* where CHEMM would be useful. Additional detail for each variable is provided by the bullets found within the Driver nodes. These bullets are also variables; they are grouped under parent nodes to simplify the structure of the model:

- *Public Health and Safety Drivers* include: the need for *Training, Planning and Preparedness* for emergencies; the need for *Personal Protection;* and the need for *Treatment of Injured* individuals.
- Drivers associated with the US Department of *Health and Human Services,* include: its *Mission and Mandate;* the resulting *Prioritization and Funding;* and the *Capability and Expertise* of HHS employees that can be applied to its activities.
- The *Types of Chemical Exposure Emergencies* relevant to CHEMM include: *Industrial Accidents; Transportation Accidents;* and *Terrorism.*

The influences of these variables propagate through the Model from one node to the next as illustrated by the arrows connecting the nodes. The arrows indicate the *primary* influences and the *primary* direction of influence. As such, other influences may exist between nodes even if no arrow is shown and some influence may exist going in the direction opposite of that indicated by the arrows, reflecting a feedback over time.

At the center of the model are the nodes representing CHEMM itself (Fig. 8.3). The *Quality of Information Content* and the *Quality of Information Provision* are depicted as the primary variables determining the ease and effectiveness with which the user can access information contained in CHEMM. Research participants in this

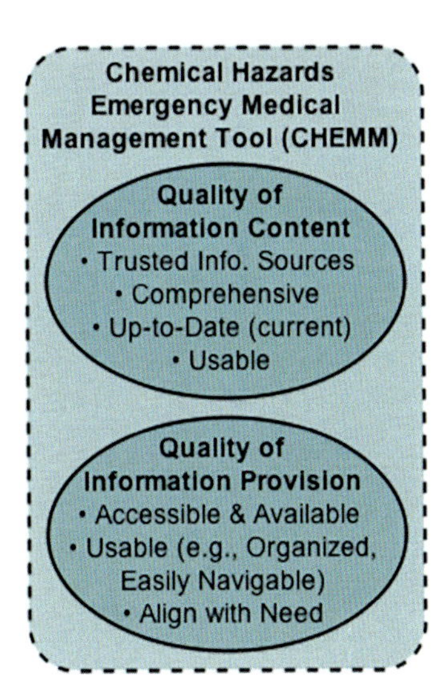


**Fig. 8.3** The CHEMM Tool

project identified a number of specific criteria (or subvariables) within each of these (Fig. 8.3):

- The *Quality of Information Content* is determined by the degree to which the information:
  - Comes from *Trusted Sources* that are authoritative, credible, validated, and credentialed, preferably from an experience-, evidence-, and consensus-based (peer reviewed) process;
  - Is *Comprehensive*, Precise, and Accurate;
  - Is *Up-to-date*; and
- Is *Usable,* with a decision-making focus and at an appropriate language level.
- The *Quality of CHEMM Information Provision* is determined by the degree to which the information is provided in a manner that is:
  - Easily *Accessible, Available,* and Reliable;
  - *Usable* (e.g., well *Organized* and *Easily Navigable*; and
  - *Aligned with Needs* of the User in a manner that is robust, customizable, and effective for a wide range of users, contexts, and entry points.

Additional variables related to CHEMM are presented in the *User Perspective* model presented on in the next section (Fig. 8.8).

The *Information Needs* node in the lower left side of the model represents the types of information needed by *CHEMM Stakeholders*—including first responders,

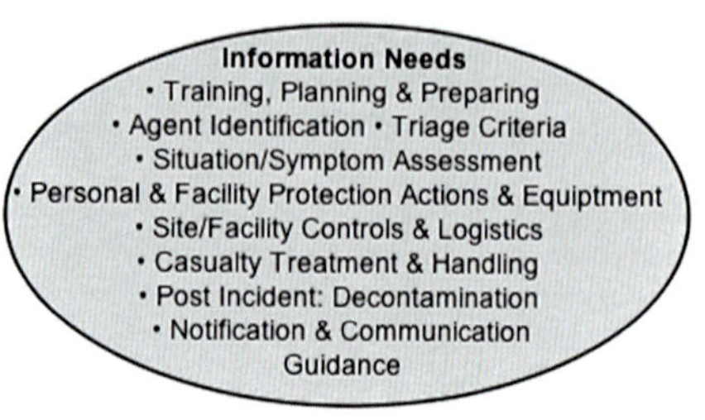


**Fig. 8.4** Information Needs

**Fig. 8.5** Quality of Stakeholder Engagement

Quality of
Stakeholder Engagement,
Strategic Partnerships & Marketing
• Information Providers
• Information Users (e.g., First
Responders & Receivers)
• Stakeholder Agencies
& Associations

first receivers, trainers, planners, or other key users—depicted in the node immediately above it. Examples of *Information Needs* include (Fig. 8.4):

- *Training, Planning, and Preparing;*
- *Agent Identification;*
- *Triage Criteria;*
- *Situation and Symptom Assessment;*
- *Personal and Facility Protection, Actions, and Equipment;*
- *Site and Facility Controls and anticipatory guidance to support Logistical decisions;*
- *Casualty Treatment and Handling;*
- *Post Incident Decontamination;* and
- *Notification and Communication Guidance.*

As illustrated by the *Quality of Stakeholder Engagement* node near the center of the model, CHEMM stakeholders can be both, information users (e.g., seeking needed information in an emergency scenario or training exercise), or information providers (e.g., providing information about experiences in emergency scenarios or feedback on the use of CHEMM, Fig. 8.5).

The *Entry Points* node depicts the initial data specified by users that is used by CHEMM to highlight options most frequently needed for the user type or informational need to help them get to the needed information quickly. The *Entry Points* include (Fig. 8.6):

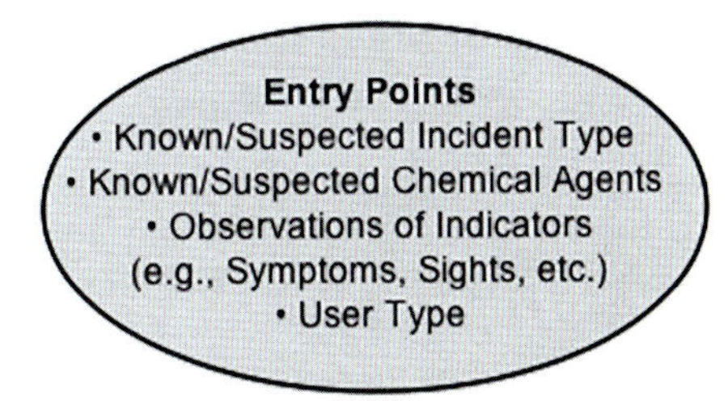


**Fig. 8.6** Entry Points

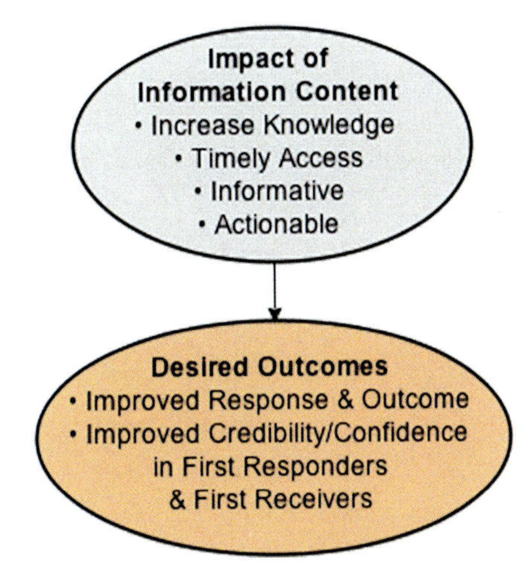


**Fig. 8.7** Impact of Information and Desired Outcomes

- *Known/suspected Incident Type;*
- *Known/Suspected Chemical Agents;*
- *Observable Indicators (for example, Symptoms, Sights, etc.);* and
- *User Type*

The model ends with the nodes in the lower right where the CHEMM *Information Provision* is shown influencing the *Impact of Information Provision*, and ultimately, the *Desired Outcomes,* for public health and safety (Fig. 8.7).

The *Impact of Information Provision* is the degree to which the information provided by CHEMM: will *Increase Knowledge*, provides *Timely Access*, is *Informative*, and is *Actionable*.

The Impact of Information Provision directly influences the *Desired Outcomes* for CHEMM use. Impacts include: *Improved Response and Outcome* in emergencies; and *Improved Credibility/Confidence in First Responders and First Receivers*.

## *Influences of the CHEMM Tool on Planning, Preparedness, and Emergency Response to Hazardous Chemical Exposures: User Perspective*

The User Perspective CHEMM Expert Model is presented in Fig. 8.8. As with the previous model, this model starts in the upper left corner with the *Drivers* or initiating influences on the system. In this model, the *Drivers* relate to the accidents and incidents where CHEMM would be useful along with associated training, planning, and preparedness activities.

The *Desired Outcomes* include (Fig. 8.9):

- *Meets Information Needs;*
- *Effective Training, Planning, and Preparedness;*
- *More Efficient Response (Less Duplication of Effort);* and
- *More Effective Response (Increased Capacity and Improved Outcomes).*

CHEMM Users include (Fig. 8.10):

- *First Receivers*
  - *ER Doctors*
  - *ER Nurses*
- *First Responders*
  - *Ambulance, EMT*
  - *Fire, Police*
  - *Hazmat Responders*
- *Other Users*
  - *Accident/Incident Coordinators*
  - *Second Responders and Receivers* (e.g., *decontamination workers, operating room or ICU staff, and mortuary personnel)*
  - *Communicators*
  - *Planners* (e.g., *those in the preparedness community)*
  - *Trainers* (e.g., *educators and trainers of emergency response and medical personnel)*

The type of user, and their responsibilities in an event, influences their initial needs and assessment of the situation. This initial assessment includes (or is influenced by) the information that has been reported to them and their own observations on the scene. This assessment, by user type, influences how CHEMM will be accessed.

CHEMM is designed to be accessed through multiplatform *Access Modes* including: desktop and laptop computers, along with handheld devices (e.g., PDAs, cell phones/smart phones)

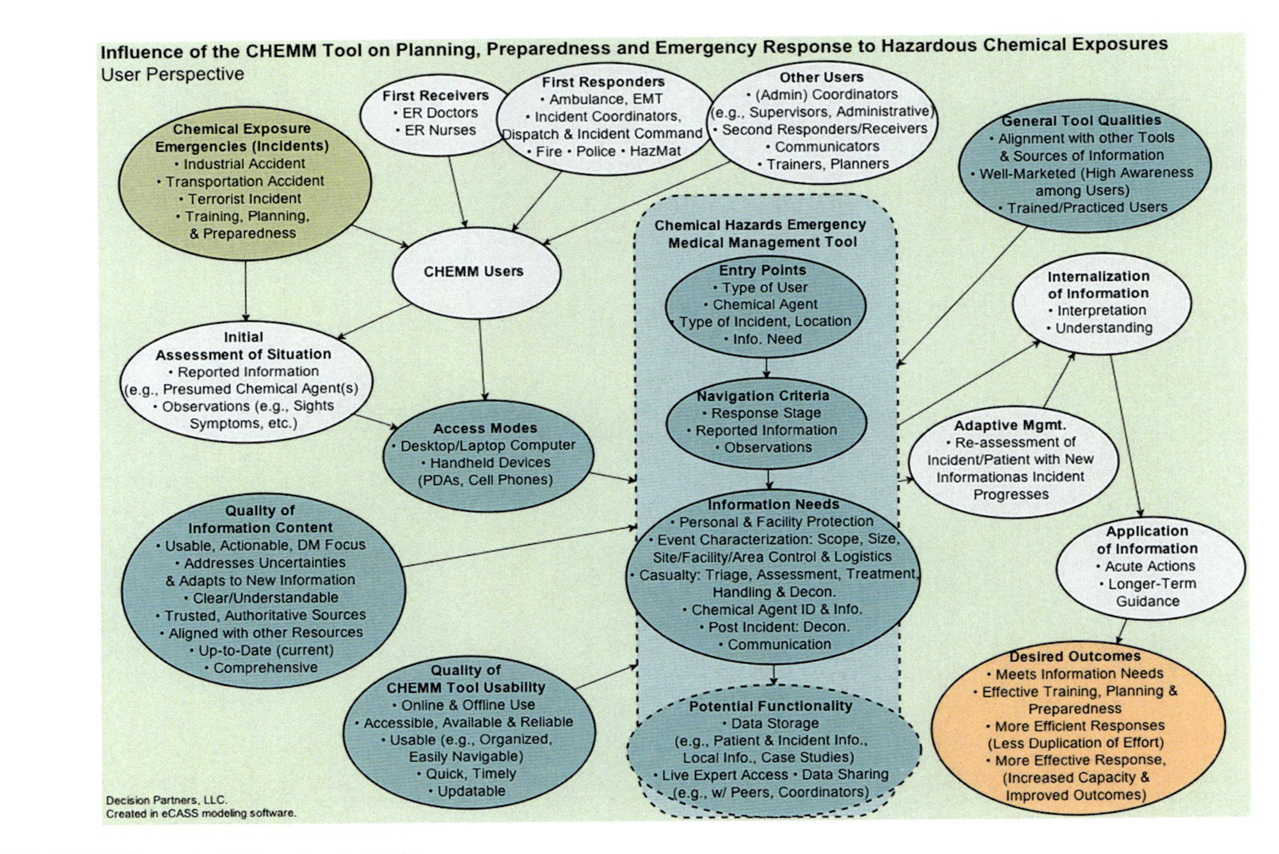


**Fig. 8.8** CHEMM expert model—user perspective

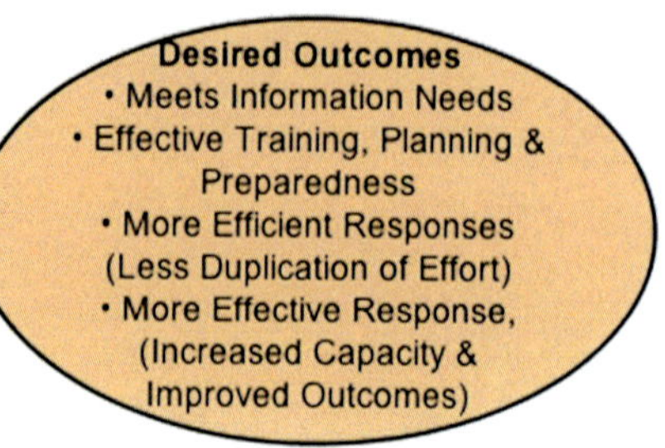


**Fig. 8.9** Desired Outcomes

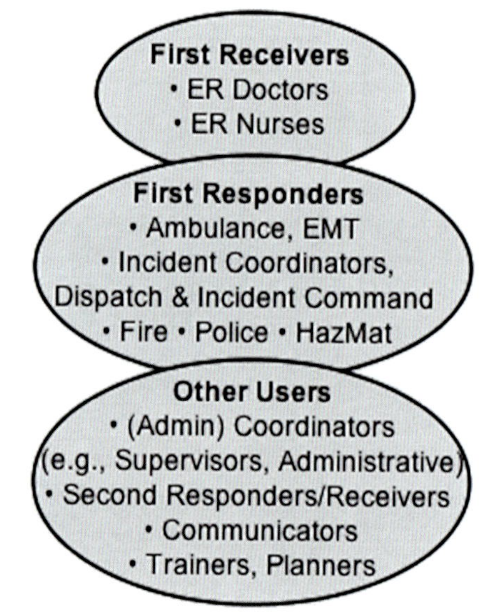


**Fig. 8.10** CHEMM Users

CHEMM itself is represented by the system of nodes found in the center of the Model. These nodes represent a process flow describing how the Tool is used.

Once accessed, the user will start the process by utilizing a particular Entry Point designed in a way that will result in the customization of the Tool interface to highlight options most frequently needed in order to optimize the user experience and help them get to the needed information quickly. The anticipated Entry Points include:

- *Type of User:* (e.g., EMT, ER Nurse, ER Doctor, etc.);
- *Chemical Agent*;

- *Type of Incident:* (e.g., terrorist incident, transportation accident, industrial facility accident); and
- *Information Need.*

After specifying the Entry Point, users will be able to navigate through the Tool by specifying Navigation Criteria, including:

- *Type of User*/Incident if not specified as the initial entry point.
- *The Stage of the Response* (e.g., preparing to go to scene, first arrival on scene);
- *Information that has been Reported to them* (e.g., type of incident, reported chemicals); and
- *User Observations* (e.g., odors, symptoms, etc.).

After specifying the *Navigation Criteria*, the user then specifies their *Information Needs*, including:

- *Personal and Facility Protection* including decontamination procedures;
- *Event Characterization* (e.g., scope, size of incident; incident site, facility, or area/perimeter control guidance to support Logistical decisions)*;*
- *Casualty Handling* (e.g., Triage, Assessment, Treatment, Handling, and Decontamination);
- *Chemical Agent Identification and Information;*
- *Post Incident Activities including Decontamination; and*
- *Communication.*

Note: The Tool interface may be able to be designed in a flexible way to allow inputs to be specified in an order other than indicated earlier (e.g., starting directly with an informational need) although such options should be tested to assess the potential impact on Tool usability (e.g., does increased flexibility add confusion and cause difficulty in training.)

- *Potential Functional Needs* of the Tool identified by workshop participants and which may be added to the Tool at a future time, include:
  - *Data Storage* (e.g., incident and/or patient information, local information, case studies);
  - *Live Expert Access;* and
  - *Data Sharing* (e.g., among other Users on scene, incident coordinators, first receivers, peers, etc.).

The overall quality of CHEMM is influenced by the nodes in the lower left: *Quality of Information Content*, which represents the quality of the information contained in and provided by the Tool, and *Quality of CHEMM Usability*, which represents the ability of the user to access that information. The overall quality is also influenced by the node in the upper right, the *General Tool Qualities*, its context in relation to other tools.

- *Quality of Information Content*:
  Workshop participants identified the following criteria as measures of the quality of information provided by the Tool, including whether the information:

- Is *Usable, Actionable* and has a *Decision-Making Focus* (e.g., time sensitive, and provides data relevant to specific time frames of incident response and predictive information about what to look out for based on current knowledge);
- Acknowledges and *Addresses Uncertainties* and *Adapts to New Information*, and provides both plausible options given current data and options for reducing uncertainty;
- Is *Clear and Understandable*, in a language level appropriate for the intended audience (e.g., Tool users and/or the lay public if the information is intended to be passed on to them);
- Comes from *Trusted, Authoritative Sources* that are credible, validated, and credentialed (e.g., experience-, evidence-, and consensus-based and peer reviewed);
- Is *Aligned with other Resources*
- Is *Up-to-Date;* and
- Is *Comprehensive,* Precise, and Accurate.

• *Quality of CHEMM Tool Usability:*
The quality of the *CHEMM Tool Usability* will be measured by the following criteria identified by research participants, including whether the information:

- Has flexible *Access Modes* rendering it readily available whether the user is connected to the Internet or is accessing an off-line database in a computer or handheld device;
- Is *Accessible, Available, and Reliable*;
- Is *Useable*, well *Organized*, and *Easily Navigable*; with readily understandable navigation options at appropriate language levels;
- Is *Aligned with Needs* and is robust, customizable, and effective for a wide range of contexts and entry points; and
- Provides data *Quickly* and *Timely* (e.g., "just in time" delivery with very low "downtime");
- Is *Updatable*, "evergreen," and is continuously improved with feedback from users and experts.

• *General Tool Qualities*:
The following are criteria, suggested by research participants, by which the quality of the Tool can be evaluated:

- *Alignment with other Tools and Sources of Information* that are designed to be used for related information needs, including:

  WISER and REMM Tools
  CDC Epidemic Information Exchange (Epi-X)
  HHS Secretary's Operations Center (SOC)
  FEMA National Incident Management System (NIMS)
  NFF Near Miss Reporting System
  Poison Control Centers
  CAMEO
  CHEMTREC

  - That it is *Well Marketed* resulting in a *High Level of Awareness* (and adoption) by the targeted *Users*; and
  - That Tool *Users are Well Trained and Practiced* on the Tool enabling them to obtain the information that they need.

The overall quality of the Tool then influences Outputs that include:

- Internalization of chemical hazards information by Tool users as measured by their correct interpretation and understanding of information provided by the Tool.
- *Application of Information*, including:
  - *Short-term (Acute) Actions*. For example:

    Preparatory checklists for first responders on the way to the scene and for receivers before the ill or wounded arrive;
    Recognizing short-term exposure indications;
    Triage and treatment of acute illness or injuries, and mental health treatment;
    Procedures for personal protection and the protection of others, and equipment needs;
    Site and area control procedures, including safety perimeters and decontamination procedures;
    Assessment and diagnostic procedures, including chemical identification and chemical information; and
    Communications, including alerting of other officials, and communications to family members, media, public, and others.

  - *Guidance on Longer Term Actions*. For example:

    Long-term treatment and mental health treatments;
    Recognizing long-term exposure indications;
    What to communicate at a later time; and
    Decontamination procedures.

Over time, Tools users may obtain new information that allows them to reassess the situation and use this new information to return to the Tool and update their actions based on new information. This is represented by the *Adaptive Management* node, which is influenced by the *Quality of the Tool* and influences the *Internalization of Information*.

The *Desired Outcomes* of the Tool's use include:

- *Meeting Tool User Information Needs*;
- *Effective Training, Planning, & Preparedness*;
- *More Efficient Response with Less Duplication of Effort*; and
- *More Effective Response*, including *Increased Capacity and Improved Outcomes*.

## User Matrices for CHEMM Optimization

Research participants also provided valuable insight into specific users and the range of situations where CHEMM would be used and the specific information needs associated with that situation. This information was particularly useful to CHEMM development as research participants highlighted the importance of customizing the information provided by CHEMM, and the manner in which that information is accessed, by the specific type of user and use scenarios. The matrices for two of the primary CHEMM users groups are presented here (Tables 8.1 and 8.2).

**Table 8.1** CHEMM user matrix of first responders

| First responders: ambulance, EMT, on-site first responders | |
|---|---|
| • Training | • Emergency response scenarios |
| • Planning | • Personnel needs<br>• Equipment needs |
| • Preparing to respond | • What to expect on scene, based on:<br>– Scale of incident<br>– Quantity released<br>– Number of potential victims<br>• Preparatory check lists<br>• Preparing for arrival on scene:<br>– Personal protection needs<br>– Equipment needs<br>– Overview of initial actions upon arrival |
| • Arrival on scene | • Checklist for initial actions, observations to be made:<br>– Personal protection checklist<br>– Equipment checklist<br>– Site/area control procedures, safety perimeters<br>– Decontamination procedures |
| • Casualty triage | • Triage instructions and criteria |
| • Casualty assessment | • Procedures for assessing, diagnosing casualties:<br>– Procedures for identifying chemical agents<br>– Information about chemical agent(s)<br>• Potential treatment options<br>• Decision-making support |
| • Casualty treatment | • Treatment guidance:<br>– Acute treatments<br>– Longer term treatment/guidance<br>– Mental health treatment/guidance<br>– Guidance for specific populations (elderly, pregnant women, pediatric patients) |
| • Casualty transport | • Transport and patient handling instructions |
| • Communicating with:<br>– Casualties<br>– Uninjured public<br>– Family members<br>– Other responders<br>– First receivers | • Communications guidance:<br>– Explanations of diagnoses, treatment to casualties<br>– Communicating/alerting of other responders/officials<br>– Prioritization of most important information<br>– Providing guidance on what to look for at a later time |

**Table 8.2** CHEMM user matrix of first receivers

| First receivers: ER doctors and nurses; on-site medical receivers | |
|---|---|
| • Training | • Training scenarios |
| • Preparing to receive | • Information about the patient status<br>• Background on the known chemical agent(s)<br>• Treatment options<br>• Patient handling information |
| • Casualty triage | • Triage instructions and criteria |
| • Casualty assessment | • Procedures for assessing, diagnosing casualties:<br>– Information about chemical agent(s)<br>– Procedures for identifying/verifying chemical agents<br>• Potential treatment options<br>• Decision-making support |
| • Treatment | • Treatment guidance:<br>– Acute treatments<br>– Longer term treatment/guidance<br>– Mental health treatment/guidance<br>– Guidance for specific populations (elderly, pregnant women, pediatric patients) |
| • Communicating with:<br>– Casualties<br>– Uninjured public<br>– Family members<br>– Other responders<br>– First receivers | • Communications guidance:<br>– Explanations of diagnoses, treatment to casualties<br>– Communicating/alerting of other responders/officials<br>– Prioritization of most important information<br>– Providing guidance on what to look for at a later time |

## Deeper Insight into CHEMM Users' Mental Models

In the next stage of the CHEMM Project, in-depth mental models research interviews were conducted with a sample of 40 First Responders, First Receivers, and Trainers/Planners. The purpose was to enhance the design of CHEMM based on insight into users' informational needs in various scenarios.

### *Research Sample*

Forty stakeholders representing a sampling of key specialty areas and geography were identified and recruited into three cohorts: First Responders (12); First Receivers (13); Planners, Trainers, and other Interested Stakeholders (e.g., Government and Academia) (15). Interviews were conducted between May and August 2011 and averaged 48 min in length.

The interviewees were generally highly trained and experienced professionals with many reporting having medical or other advanced degrees. Those who spontaneously mentioned their years of work experience reported having 6–36 years of experience, with an average of over 20 years. Their education, training, and experience often cut across cohort boundaries, providing excellent representation of the various perspectives. Interviewees were also geographically dispersed throughout the United

States, located and working in both urban and rural areas across 16 states and Washington DC.

Places of employment and professional titles of the interviewees across the three cohorts spanned a broad range, including:

**First Responders**

- **Employment:** local fire departments, private association, state agency, military, FEMA, Native American Tribal Nation
- **Titles:** Firefighter, EMT, Paramedic, Captain, Operations Manager, Safety Officer, Incident Commander, Toxicologist

**First Receivers**

- **Employment:** hospitals, federal government, private corporation
- **Titles:** Director of Emergency Medical Services; Paramedic, Emergency Physician, Nurse, Hospital EMS Communications Manager, Hospital Emergency Preparedness Coordinator, Hospital Safety Officer

**Planners, Trainers, and Others**

- **Employment:** local fire departments, private company, poison control center, hospital, Homeland Security/FEMA, military, other federal agencies, state or local agencies, academia, professional associations
- **Titles:** Emergency Manager, Program Manager, President, Vice President, Education Coordinator, Trainer, Hazardous Material Officer, Medical Officer, Director, Deputy Chief, Bioterrorism Trainer, Toxicologist, Industrial Hygienist, Communication Duty Officer, Nurse, Research Scientist

In many cases, interviewees were cross-trained and/or employed in more than one capacity, for example, an emergency room physician trained as a firefighter, or an individual who runs a disaster response command and control center who also works as a part-time nurse and volunteer firefighter. Due to the robust, interdisciplinary backgrounds of the Interviewees, detailed analysis, comparisons, and contrasts between the cohorts was impracticable as interviewees often provided different perspectives reflecting the different roles in their responses to our questions.

## *Interview Topics*

Interviewees were offered an agenda of topics to discuss in a way that allowed for free expression and they were encouraged to raise additional topics spontaneously and to elaborate on their perspectives. Interview topics included:

- Tool Uses and Information Needs: Potential uses of CHEMM and associated information needs that could be addressed by CHEMM.
- Information Quality Criteria: Criteria that NLM or users would use to evaluate the quality of information provided by CHEMM.

- Features and Functional Criteria: Criteria that NLM or users would use to evaluate CHEMM's functions and features.
- Engaging CHEMM Users: Engaging and marketing to potential CHEMM users.
- Marketing and Continuing Development of CHEMM.

## Summary Mental Models Research Findings

Interviewees expressed a high level of enthusiasm and interest in the potential of the CHEMM Tool and were very complimentary about the research process commenting that they appreciated the opportunity to provide input. The research results were consistent with the February 2010 Stakeholder Workshop results, but added a richness and depth of knowledge that was critical for optimal development.

**Top-Line Findings**

- The types of information sought by users (e.g., personal protective equipment, chemical information, casualty assessment and treatment, etc.) are relatively consistent across all user types suggesting potential implications for CHEMM design.
- Most interviewees, across all cohorts, perceived "trust" as the primary determinant of information quality, using terms such as: "authoritative," "reputable," "credible," "vetted," "peer-reviewed," "accurate," "evidence-based," and "verified."
- Some highlighted the importance of "usability," expressing a need for clear information that is quickly accessible, concise, "straight to the point," and delivered by an "easy to use" platform.
- Some also focused on "alignment" with the numerous resources already available, providing a "one-stop shop" for trusted and credible information.
- Interviewees stated that CHEMM could and should be used in many contexts, not just mass casualty incidents. Additional and more frequent uses will help users maintain familiarity with and competence in CHEMM use and increase the likelihood that it will be used, and used effectively, during a mass casualty event.
- Drawing on their experience of working with many products that do not live up to expectations, interviewees recommended a continuous program of engaging users and evaluating the Tool's effectiveness.
- Interviewees believed that by providing information specifically selected and organized by the user, CHEMM would be "incredibly helpful." Additionally, filling information gaps would make them feel "more comfortable making a decision" in "high stakes" situations.
- Many interviewees anticipated some functional issues that would have to be overcome, including memory limitations for handheld devices, connectivity and bandwidth for Internet connected devices, updating issues for off-line devices and even government restrictions on types of devices that can be purchased.

## *CHEMM Uses and Information Needs*

When asked about the types of incidents where access to an informational tool would be useful, Interviewees primarily mentioned the anticipated incident types discussed in the February 2010 Workshop, including: Chemical accidents associated with transportation and at industrial facilities; and intentional releases from terrorism incidents. Some also mentioned additional incident types, not raised in the previous research steps, such as chemical accidents at water treatment facilities, agricultural facilities, military facilities, and unconventional locations such meth labs.

Interviewees mentioned many types of information that they would be looking for in an incident, most of which had been identified in the earlier steps of the CHEMM research, including (in descending frequency of mention):

- Identification and characterization of chemical agents (e.g., toxicology, chemical, and physical properties)
- Symptoms, triage, assessment, and treatment of those affected
- Type and use of appropriate personal protective equipment to be used in an incident
- Perimeter and incident site control
- Decontamination of people and site
- Communications guidance

They also identified other types of information, not raised in previous research steps, including:

- Location-based information about the area or facility where an incident occurs, local mapping, resources available local to incident, distance receiving care
- Incident characterization (e.g., typical magnitude and scope of this type of incident, and available resources)
- Plume predictions, hazard zones, and casualty predictions
- Definitions, glossaries of terms
- Case studies for training purposes
- Environmental protection measures
- Veterinary (e.g., livestock) risks

Interviewees reported that they currently obtain information from a number of government and nongovernment sources, including:

| Government Agencies | Nongovernmental and Industry Organizations |
|---|---|
| • ATSDR | • Poison Control Centers |
| • CDC | • Weapons of Mass Destruction Civil Support Teams (WMD-CSTs) |
| • DHS | • Interstate Chemical Threats Workgroup (ICTW) |
| • DOD | • American Industrial Hygiene Association (AIHA) |
| • DOT | • American Conference of Governmental Industrial Hygienists (ACGIH) |

(continued)

| Government Agencies | Nongovernmental and Industry Organizations |
|---|---|
| • EPA | • Association of Public Health Laboratories (APHL) |
| • FBI | • National Fire Protection Association (NFPA) |
| • FDA | • CHEMTREC |
| • FEMA | |
| • NIOSH | |
| • OSHA | |

They mentioned a number of specific informational resources provided by government and nongovernment agencies, including databases, guidebooks, handbooks such as the Emergency Response Guidebook (ERG, also called DOT or Orange Response Guidebook), the NLM Hazardous Substances Data Bank (HSDB), Material Safety Data Sheets (MSDS), the NIOSH Pocket Guide, and even Google.

They also discussed a number of electronic platforms for chemical and emergency response information and incident management platforms, including CHEMM's sister applications also produced by NLM (WISER—Wireless Information System for Emergency Responders, and REMM—Radiation Emergency Medical Management), and other platforms such as CAMEO (Computer-Aided Management of Emergency Operations), Poisindex and EPI-X (Epidemic Information Exchange).

When asked what information was lacking in the currently available resources, Interviewees identified a number of gaps. Many related to daily incident response requirements, but some related specifically to rare mass casualty events. Examples of lacking information included:

- Personal protective equipment appropriate to the incident.
- Definitive guidance to help resolve the problem of "five different answers from three sources."
- Chemical reactions and multichemical interactions.
- Exotic chemicals that aren't in other databases.
- Mass casualty information and how these events differ from isolated incidents.
- More detailed medical treatment information.
- Decontamination information for equipment and waste disposal guidelines.

Some commented that the greatest need wasn't for new information, but to have that information available in one, easily accessible and usable resource.

After hearing a description of CHEMM, Interviewees' initial reactions were almost universally positive, sometimes enthusiastically so, noting the need and potentially strong demand for such a tool, and specifically highlighting the potential benefits of CHEMM, pulling together information from multiple sources into one resource that can be utilized by different types of users for multiple use scenarios, including incidents of all sizes, not just mass casualty incidents.

The description of CHEMM and subsequent discussions raised questions for some, such as how its design will be user friendly and intuitive to meet multiple users' immediate needs upon arriving at a scene, and how it will align with other resources already available, including WISER. A few questioned the primary function of CHEMM—an informational tool, a decision support tool, or a decision-making tool—suggesting the need to clearly define and communicate its purpose to potential users.

When discussing how they might use CHEMM, Interviewees mentioned a wide range of uses for specific user type, suggesting that First Responders would use CHEMM for guidance on personal protection equipment and for support on assessment, treatment, and handling of casualties; First Receivers would use CHEMM for support on chemical identification, triage, assessment, and treatment of patients; and Planners and Trainers would use CHEMM as a resource to study chemicals and develop planning and training scenarios.

## *CHEMM Information Quality Criteria*

When asked what terms Interviewees would associate with high quality and authoritative information, their responses were similar to the terms defined in the workshop, though the larger number of individuals in the research provided an opportunity to qualitatively rank the importance of these qualities. (Note: due to the small sample size, this is not statistically representative of the overall stakeholder population) (Fig. 8.11).

Many interviewees suggested that NLM could ensure that information provided by CHEMM would be of high quality, by requiring that the information comes from vetted source and that it is peer reviewed and verified by federal, state agencies, and other organizations on a continual basis and through user testing with those in the field.

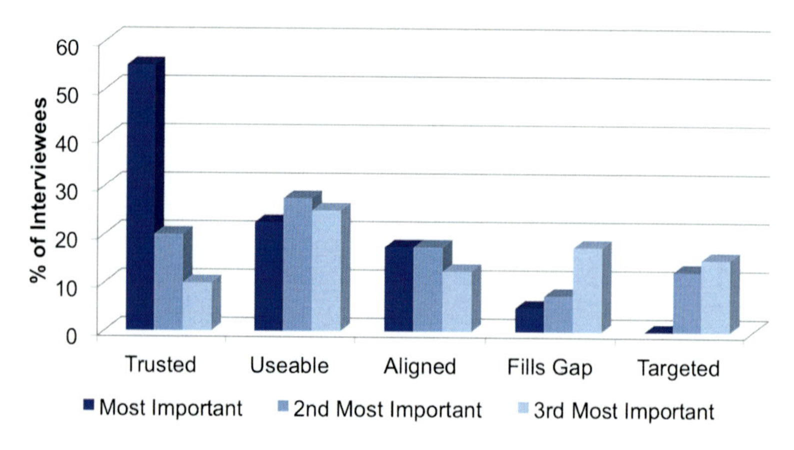


**Fig. 8.11** Ranking of quality criteria preference

## *CHEMM Functionality*

Interviewees across all cohorts indicated that they would most likely access CHEMM primarily using a laptop or a handheld device, although many also said they would access it using multiple platforms (e.g., desktop, laptop, or handheld device), depending on the situation and the location they would be using it. Most wanted to be able to access CHEMM both online and offline depending on the circumstances, but they said they generally prefer to work online whenever possible.

Many anticipated challenges to connectivity related to: power and connectivity outages that may be associated with certain types of incidents; bandwidth and memory limitations for mobile handheld devices; Government restrictions on the types of devices that government employees are allow to use; and ensuring that databases are kept up-to-date, especially for devices that do not have a constant Internet connection.

Thinking about the types of data that they would input into CHEMM to get at the information that they needed, most interviewees said they would enter the chemical agent if it is known, or go through an identification process if it is not known. They said they might also specify the type of user that they are (first responder, first receiver, or more specific job), or information about the type of incident and the stage of the response. Interviewees discussed several other features and types of functionality that they believed would add value to CHEMM including:

- Recording and printing a log of actions taken for future reference for training and legal purposes.
- Providing access to live experts for immediate guidance from sources such as Poison Control or CHEMTREC.
- Integrating system connected to incident management systems.
- Networking systems that share data with other responders, receivers, and incident command to improve coordination and reduce duplication of effort.
- Enabling the use of pictures for diagnosis, documentation, or sharing information during an event.
- Enabling GIS-based functionality to locate patients, site features, etc., or to document conditions.
- Offering a customizable interface that allows CHEMM users to more quickly access the specific information they routinely need.
- Allowing access to locally available information (e.g., local facilities) including data input by CHEMM users (e.g., during planning or training.)

## *Stakeholder Engagement and Continuing CHEMM Development*

When asked if there were other potential users or individuals with whom NLM developers should engage as they continued to develop CHEMM, Interviewees suggested a detailed list of potential users, including: other Federal, State, and Local

Government agencies; individuals from industry (chemical producers, transporters, and users); Professional Associations that represent stakeholders; additional responders, receivers, planners, and trainers; with some suggesting the general public

Interviewees suggested a number of approaches that NLM could take to make people aware of and familiar with CHEMM, such as promoting it at conferences and/or through professional associations affiliated with the responder and receiver users. Many indicated that conducting interviews similar to the one in which they participated, along with similar ongoing direct outreach with individuals to get feedback was a good approach for making people aware of CHEMM. Many suggested pilot or beta-testing in "real-world scenarios" with real users was critical. Some suggested forming strategic partnerships with other agencies, associations, and hospitals including integrating into existing training programs and using federal funding and grant mechanisms to encourage CHEMM use. Many even suggested using print advertising, email, and the Internet to promote CHEMM.

### *Interview Wrap-Up and Interviewees' Closing Thoughts*

At the close of the interview, interviewees provided thoughts and advice on how to ensure the quality and effectiveness of CHEMM. They recommended the CHEMM Team commit to a sustained program of getting user input, testing, and evaluating effectiveness, and that they keep making refinements after CHEMM is on the market. They also reiterated the need to ensuring that CHEMM uses trusted, credible information, and that it is designed to be simple to use given the high-pressure nature of the scenarios where it will be used. They also suggested developing strategic partnerships in the stakeholder and user community to continue promotion and obtaining real-world feedback. Some reiterated the need to clearly communicate about CHEMM potential uses and to consider broader uses and features for CHEMM.

## Building on the Mental Models Research Results

Following the mental models research, a CHEMM Strategy Development Workshop was held in Washington, DC, on December 14, 2011 with key CHEMM stakeholders and collaborators update them on the results of the mental models research and to define and discuss strategies and prioritize objectives for CHEMM's development going forward.

For the near term, workshop participants recommended maintaining focus on CHEMM's primary mission and primary users, while expanding outreach to users to continue to gain insight into how the tool can be used and what content is critical. They also suggested identifying potential partnerships to enable continued CHEMM development and building networks of users to promote CHEMM. They discussed opportunities to raise awareness among potential CHEMM users by promoting it through various channels such as YouTube, social media, NLM/NIH publications, academic literature, and through other government agencies and through the network of First Responder and First Receiver associations. They suggested partnering with those who offer emergency response training to raise awareness, understanding, and use of CHEMM. They also highlighted the need to align CHEMM development and offerings with that of other Emergency Response tools to develop a suite of tools to address chemical, biological, radiological, nuclear, and explosive (CBRNE) emergencies.

Finally, the results of the CHEMM research were used to develop a comprehensive Stakeholder Outreach and Communications Strategy and Plan for 2012–2014 for the NLM Suite of CBRNE Emergency Preparedness and Response Tools, focusing on the common stakeholders and interconnected objectives related to NLM's broader suite of tools targeted at CBRNE threats. This Plan offers specific considerations based on stakeholders' needs, interests, and priorities, discovered in the CHEMM research. The Plan includes an overall strategy, an overview of specific communications materials needed, and guidance for a science-based approach to outreach and communications with stakeholders critical to effective emergency preparedness and response.

## Client Perspectives on Mental Models Research, Key Learnings, and Applying the Results

When asked to reflect on the purpose, outcomes, and application of results from the mental models project, the NLM client lead for the CHEMM Project, Bert (Pertti) Hakkinen, Ph.D., highlighted the need for an exploratory process that could provide a sound foundation for a project being developed from scratch. "The opportunity we had with CHEMM was to design it from the ground up, to meet the needs of first responders, first receivers and other likely users. To do so, we needed to understand who the likely stakeholders are, what information they want and in what order." He stressed the importance of doing research to ensure that CHEMM was a usable tool, not just an academic exercise. "We're trying to develop a resource. It's not a book, or a journal. It's something where people will 'click, click, click' to get to the information they want, ideally in as few clicks as possible. ... If there's a chemical incident out on the highway, how would they deal with that? What might an individual user's process be for seeking information compared to that of somebody else?"

Dr. Hakkinen spoke of Mental Modeling as a proven approach to help the Project Team elicit the insight needed to focus CHEMM development and communications strategies. "I've been following the psychology behind risk for years, and years, and mental models was the accepted way that I knew was out there. It's been accepted by the academic community, the FDA and others. So, I knew it was an accepted and highly regarded approach. My thinking was that Mental Modeling, with expert models was an approach that should be used for this type of project."

According to Dr. Hakkinen, NLM found the Mental Modeling approach to be effective and efficient. "The mental models interviews and the mental models-based workshops were a very effective way to quickly get our initial concept looked at, in depth, by various key stakeholders. I've never been in a meeting that had this spectrum or diversity of people who were all interested in a topic like this, working with each other and talking things over. So, it was a very effective way to get to these stakeholders together to discuss a project like this where they all had a common interest. I don't know if we can say we got consensus in the workshop, but we sure got good feedback from people and nobody was really negative."

In particular, the NLM Team focused on the value of providing a framework for understanding and communicating about the ongoing challenges and focus of CHEMM development. "It helped us focus our thinking in terms of how CHEMM was structured, what people would expect in terms of content and what information would they need at certain time points during an incident. The Expert Model breaks down the system perspective into the types of exposure, the drivers, quality of information content, all of that leads to, 'Have we done this? Did we consider that?' This really helped us. It helped us then. It helps me now to think about where we are in the whole spectrum of working on the overall system perspective. This type of Mental Models thinking is always in the back of my mind for how we're thinking about CHEMM and where we go next."

Looking forward, Dr. Hakkinen saw the products of the Mental Modeling research as a solid foundation for educating others about CHEMM and helping to justify and support continuing its development. "The Expert Model figures themselves can be used as presentations. It's straightforward to walk people through the drivers, the types of incidents where CHEMM could be used, down to the lower right where you have the desired outcomes. It provides support for the scope of the project in terms of what we think should be done. It takes away some of the mystery behind how to do it or what to do by showing the whole approach and saying, 'this is from a science-based approach to identifying CHEMM content and the users' thinking behind it.'"

NLM used the results of the mental models research to adapt the initial CHEMM design to user needs, making it a more efficient and effective resource for its users. In addition, Dr. Hakkinen noted: "One example of how we have used the results of the research would be the marriage of CHEMM and WISER (Wireless System for Emergency Responders) as a single resource to help first responders and others so they don't have to have two apps. You can picture WISER and CHEMM both being in the Expert Model, even though our focus going into the project was solely on CHEMM. WISER fits in under the Industrial Accidents, Transportation Accidents nodes and this helped us in our thinking, in-house, about WISER and CHEMM as a single, broader resource."

Overall, Dr. Hakkinen spoke favorably of the benefit of the Mental Modeling approach in improving CHEMM. "We know going in that CHEMM was going to be a very complex resource. I like to say that CHEMM is like a home screen with thousands of pages of content behind it. The results of this approach help people get from the home screen to a particular page or subsection of content as easily and as quickly as possible. We probably would have done a good job on CHEMM without Mental Models, but this helped us confirm our thinking and direction and go beyond developing just a 'good' resource but a 'very good', or forgive me for saying, but an 'excellent' resource. We're now talking CHEMM 2.0 instead of CHEMM 1.0. Now we're thinking about what can we do down the road."

Finally, Dr. Hakkinen, CHEMM's Managing Editor, received the 2011 Risk Communication Award from the Alliance for Chemical Safety and the 2014 Fellow award from the Society for Risk Analysis for CHEMM.[2]

[2] See: https://chemm.nlm.nih.gov/about.htm#award.

# Chapter 9
# The Chamber of Mines of South Africa Leading Practice Adoption System

**John Stewart and Gordon Butte**

## Background to Development of the System

High accident rates have been a challenge in deep level mining operations in South Africa from as early as 1900. Much has been done since then to address these many risks and although major improvements in safety performance have been achieved, injury and fatality rates have lagged behind those of other leading mining countries. The focus on mine health and safety in South Africa was dramatically increased by preparation of input into a judicial commission into mine health and safety in 1994, by participation in the process to establish the International Labour Organization (ILO) Mine Safety and Health Convention 176 of 1995, and then by tripartite participation in the establishment of the South African Mine Health and Safety Act 29 of 1996.

One of the requirements of the new Act is that the newly established Mine Health and Safety Council must hold a Tripartite Summit (Labor, Employers and Government) at least once every 2 years to review the state of health and safety at mines. With a view to further focusing attention and effort on bringing health and safety performance in line with that of leading mining countries, a set of health and safety targets and milestones was agreed to by the tripartite partners at the Summit held in 2003. In essence, the milestones to be achieved by 2013 required about a 20 % year-on-year reduction in mine fatality rates over the intervening 10-year period. Target milestones for performance improvement in the area of health were set for both 2008 and 2013. Agreement to the milestones clearly pointed to the need for the mining industry to do something quite different from what had gone before.

J. Stewart, Ph.D.
J M Stewart Consulting, 2 Mayer Street, Johannesburg 2192, South Africa
e-mail: stewart@jmsconsulting.co.za

G. Butte (✉)
Decision Partners LLC, Suite 200, 313 East Carson Street, Pittsburgh, PA 15217, USA
e-mail: gbutte@decisionpartners.com



M.D. Wood et al., *Mental Modeling Approach*, Risk, Systems and Decisions,
DOI 10.1007/978-1-4939-6616-5_9

## Leadership Commitment and Exploratory Work

Given the enormous challenge of meeting the 2013 milestones, the members of the Chamber of Mines of South Africa recognized the need for collaboration in addressing the challenge. Chamber member mines are responsible for about 85 % of South Africa's mineral production, and employ about 85 % of the people who work at South African mines. The first important step was taken in 2005 when the CEOs of the Chamber member companies publically signed a document explicitly declaring their commitment to achieve the Tripartite-agreed milestones. This explicit commitment led to establishment within the Chamber of a Mine Health and Safety Task Force to investigate ways in which the challenge might be best addressed.

The newly established Task Force set up four focus groups to identify how best to achieve the massive health and safety improvements required to meet the 2013 milestones. Unsurprisingly, given the mining industry's historically heavy investment in research and development, the focus of one of the groups was on how best to accomplish health and safety improvements through the transfer and adoption of technology that had been developed but not yet widely adopted. There were essentially two parallel strands to the work that was undertaken by this group.

The one strand was devoted to developing an understanding of the mental models of the key people in industry involved in technology transfer processes, and the other was to review past and current experience in technology transfer in South African mines, but also more broadly. Inclusion of the strand on mental models research was largely as a consequence of the success achieved by Decision Partners in using a Mental Modeling approach to better address stakeholder judgment, decision making and behavior on sustainability in assisting the Mining Association of Canada in its industry-wide initiative known as "Towards Sustainable Mining." The review of technology transfer experience sought to learn from the hard-earned experience in implementing the research findings of the Chamber's Research Organisation, as well as to tap into more recent experience by major South African companies in achieving successful technology transfer.

Based on the findings from these two strands of investigation, a proposal of how best to address the issue of technology transfer and adoption was then developed.

## Key Outcomes

The above work was commissioned by the Focus Group on Technology early in 2007 and completed by June of that year, when it was presented to senior executives of Chamber member companies. It was agreed that the proposed system should be piloted in 2008, but that in doing so the system should be confined initially to facilitating the widespread adoption of *identified leading practices*. By the end of 2008, Chamber members agreed that implementation of what had become known as the Leading Practice Adoption System should go ahead for a further 5 years.

In regard to the Adoption System, it is important to recognize that development of much of the operational detail of the System occurred following initiation of the piloting phase. However, the System, and all of its operational developments have been documented in the form of a fully comprehensive guidance handbook (Stewart, 2014), which is used by the specialist secretariat and the adoption team managers in consistently implementing the system across the network of member mines. Some of the key aspects of both the original and later developed System are outlined as follows:

1. Expert model on innovation.

   Fundamental to development of the Adoption System was the use of an expert model of the innovation process. It was recognized that the identification, communication and operational adoption of a technology or leading practice are indeed the fundamental aspects of any innovation process. The potential benefit of applying the innovation model in the development of the Adoption System was thus clear. The key elements of the *Expert Model of Innovation* are presented schematically in Fig. 9.1.

   Although presented here largely as a linear process, it is important to recognize that the various steps may in practice overlap and require some backwards iteration to achieve the required outcomes. This simple diagram proved to be a most helpful framework for developing and presenting a systematically sound leading practice Adoption System.
2. Enabling leadership and management.

   The *Expert Model on Innovation* highlights the upfront need for a management orientation appropriate to the nature of the challenge and key tasks involved

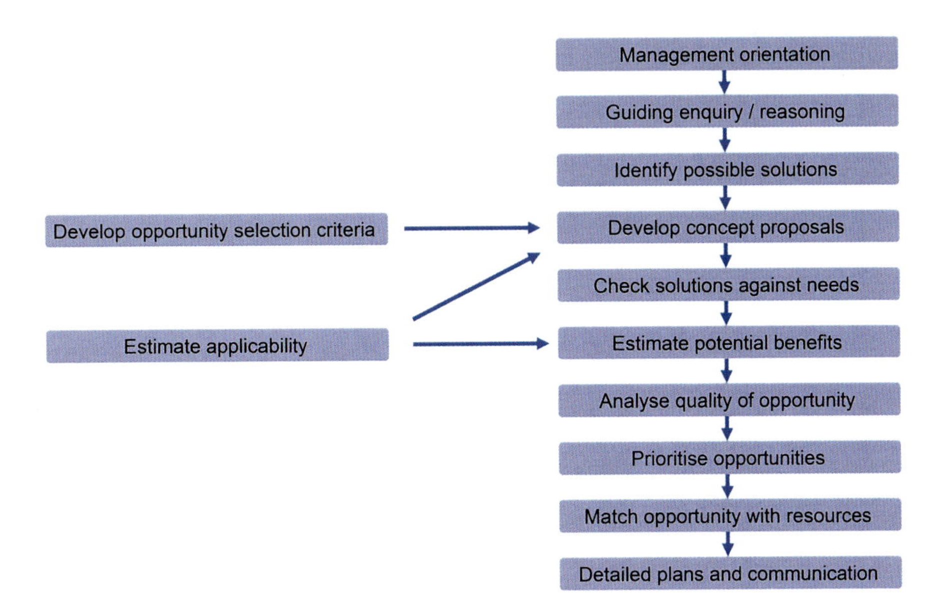


**Fig. 9.1** Expert model of innovation

in accomplishing dramatic change in occupational health and safety performance. In particular, leaders and management need to create an environment that is enabling and reinforcing of adoption activity, and personally rewarding to those engaged in the adoption process. Given that the Adoption System is for application across the various Chamber members that operate as independent mining companies, it was recognized that this orientation was needed at an industry level for the adoption processes to be effective. The Chief Executives of the Chamber member mining companies have gone a long way to achieving this by agreeing to a set of values and action-focusing operational principles in the area of occupational health and safety, and to jointly work toward the achievement of zero harm operational environments in their mines.

3. Industry ownership through direct involvement.

   The importance of achieving industry ownership and buy-in emerged as one of the most strongly and consistently expressed views in the exploratory work. Consideration of this reality led to using industry staffed adoption teams as the primary facilitating agent of the adoption process. In various ways, and at key times, these teams are supported by appropriately selected industry experts. This is a core operational principle of the Adoption System. It is particularly relevant during the planning stages, when attention is focused on the identification of key risks and potential leading practices, and when potential adoption mines are being identified.

4. Specialist secretarial support.

   Another key finding of the exploratory work was the difficulty industry persons experience in devoting time to projects that are additional to their normal operational responsibilities. With a view to easing this burden, but also with a view to ensuring specialist support and the creation of institutional memory, the system included the establishment of a specialist secretariat to support the industry adoption teams. Such specialist support is also required to facilitate and help ensure that the more complex of the various adoption processes is fully and properly undertaken. Without such support the system is unlikely to achieve its full potential.

5. Communities of practice for adoption.

   Establishing a community of practice (COPA) to facilitate the widespread adoption of a leading practice has proved to be a core aspect of the adoption process. The identification of potential adoption mines, and of the people who will play a key role in deciding and enabling adoption of the practice at those mines, is a major challenge addressed by adoption teams from an early stage of the process. Bringing such people together in a community of practice for adoption once the leading practice has been properly documented is a key determinant of the success of the adoption process.

6. Mental modeling.

   In the originally proposed system the importance of ensuring quality communication was recognized, and the need for including a focus on behavioral communication was identified. However, the full implication of this was not recognized until later, and a great deal of time and effort has been devoted to

**Table 9.1** Embedded Mental Models Processes

| Process |
|---|
| **Develop causal chain expert model of the risk situation**<br>Serves as an expert datum in developing a behavioral communications plan |
| **Identify deciders, adopters and stakeholders**<br>Specific to the selected leading practice |
| **Conduct direct enquiries with deciders, adopters and stakeholders**<br>Using a selected sample across the identified adoption mines |
| **Develop broad-based mental model**<br>Using data from the direct enquiry process across all adoption mines |
| **Develop a broad-based behavioral communication plan**<br>The plan would be broadly applicable to all adoption mines |
| **Identify desired leadership behaviors**<br>Based on detailed description of the selected leading practice |
| **Identify behavioral elements at the source mine**<br>Part of investigation to document the leading practice at the source mine |
| **Finalize broad-based behavioral plans for the leading practice**<br>Introduce identified elements into the broad-based behavioral plans |
| **Customize behavioral plans for the lead adopter mine**<br>Based on results from a direct enquiry process at the lead adopter mine |
| **Refine broad-based behavioral plans for widespread adoption**<br>For inclusion in the leading practice adoption guide |
| **Each adoption mine customizes behavioral plans for implementation**<br>Based on results from a direct enquiry process at the adoption mine |

developing the complete suite of Mental Modeling processes that now form a core and integral part of the Adoption System. The steps of the systematic Mental Modeling process employed in the Adoption System are summarized in Table 9.1. The process starts with the development of a quality expert model to properly understand the risk situation and includes a direct enquiry process involving key stakeholders to obtain input for determining the prevailing mental models. Systematic analysis then leads to the development of broadly generic behavioral communications and leadership behavior plans, which are then customized at each of the adoption mines. In executing the process it is important to recognize the interdependence of the various steps and to ensure that each step is properly executed in order to achieve an effective outcome.

## A Residual Communication Challenge

Since 2008 the Chamber's Leading Practice Adoption System has been used to achieve widespread adoption of a number of leading practices, and hundreds of people have participated in the various COPAs that have been established as part of the facilitation process. Notwithstanding this, and the success achieved, it has become apparent that many key people on mines have not yet developed a full understanding of the Adoption System and, in particular, of the Mental Modeling processes that form a key part of the adoption process.

Without this understanding, there is a temptation to take short cuts and to leave out aspects of the Mental Modeling process. This has potential for introducing long-term counterproductive consequences into the adoption process, since it is the behavioral communication and leadership behavior aspects of the adoption process that help ensure eager and sustainable adoption, as opposed to potentially short-term reluctant compliance with management dictates.

The important residual challenge is thus to make key people more aware of the full scope of the Adoption System and, in particular, of the need to ensure that all of its applicable elements are properly executed in the operational adoption of a leading practice. In order to achieve this, it was decided that a relatively brief document should be prepared to outline the key elements of the Chamber's Leading Practice Adoption System in the form of a largely pictorial brochure. A copy of this brochure is presented as an Appendix to provide a detailed overview of the Chamber's Leading Practice Adoption System.

## The Road Ahead

Logically, the road ahead will first focus on using the brochure as a strategic communication aid to facilitate engagement with key people on mines to address any uncertainties or queries they may have about the Adoption System and its various processes. In line with the concept of behavioral communications, it is envisaged that this communication process will facilitate improved application of the Adoption System processes at mines.

Importantly, attention is also being given to using the Adoption System to facilitate the adoption of management practices by mines that embrace key generic elements of the mental models processes embedded in the Adoption System. These key elements are currently being identified and packaged as a generically applicable leading practice. The objective of this initiative is to establish a leadership culture across the mining industry that will not only be broadly beneficial to mine management, but which should also enable full application of the adoption processes of the Leading Practice Adoption System.

**Acknowledgments** Special thanks is given to the Chamber of Mines of South Africa, for permission to publish the brochure on the MOSH Leading Practice Adoption System, and for providing JM Stewart Consulting and Decision Partners, and the authors, with support and freedom to collaborate closely in development of the Adoption System.

## Appendix

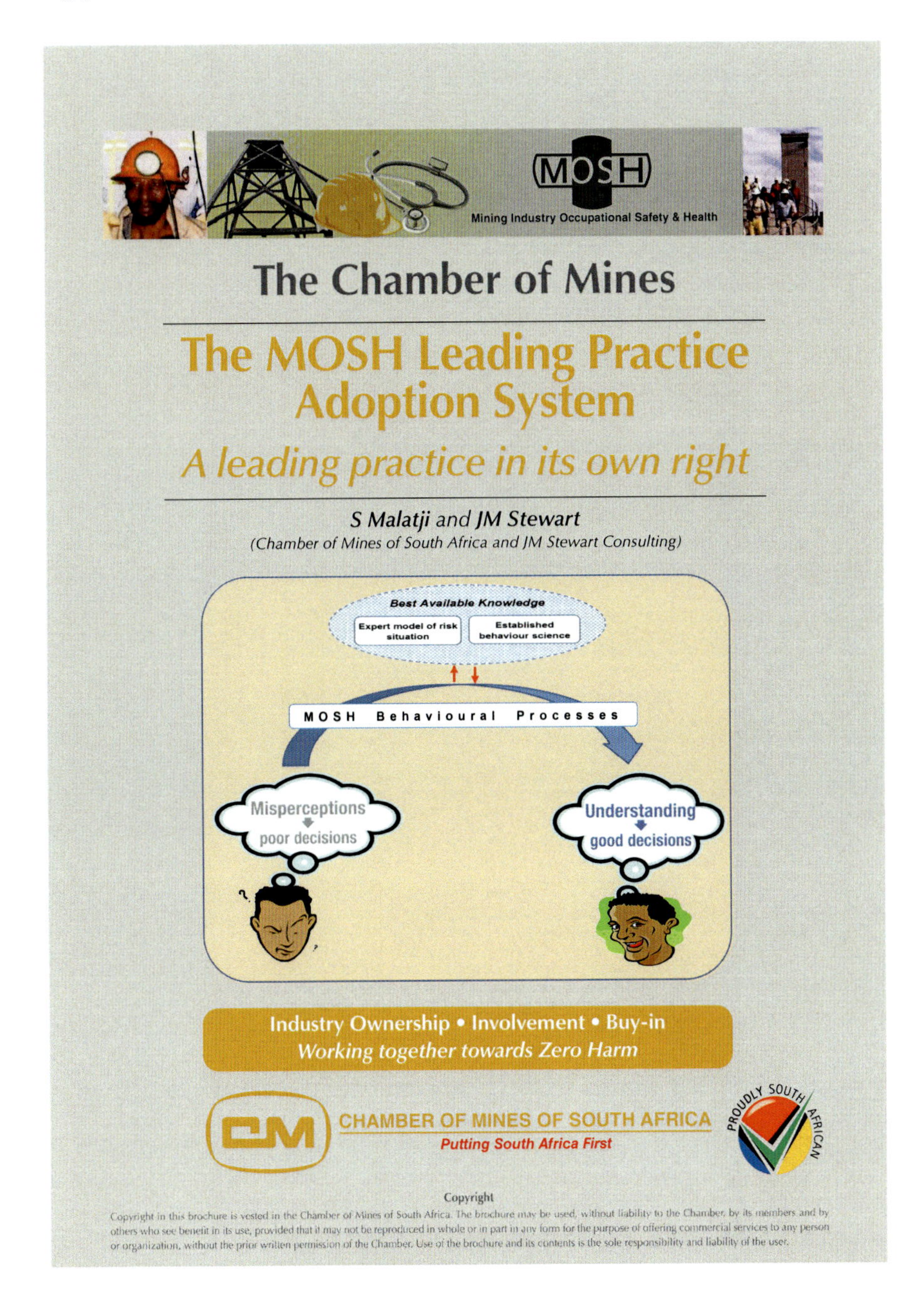
MOSH
Mining Industry Occupational Safety & Health
The Chamber of Mines
The MOSH Leading Practice Adoption System
A leading practice in its own right
S Malatji and JM Stewart
(Chamber of Mines of South Africa and JM Stewart Consulting)
Best Available Knowledge
Expert model of risk situation
Established behaviour science
MOSH Behavioural Processes
Misperceptions
poor decisions
Understanding
good decisions
Industry Ownership • Involvement • Buy-in
Working together towards Zero Harm
CM
CHAMBER OF MINES OF SOUTH AFRICA
Putting South Africa First
PROUDLY SOUTH AFRICAN
Copyright
Copyright in this brochure is vested in the Chamber of Mines of South Africa. The brochure may be used, without liability to the Chamber, by its members and by others who see benefit in its use, provided that it may not be reproduced in whole or in part in any form for the purpose of offering commercial services to any person or organization, without the prior written permission of the Chamber. Use of the brochure and its contents is the sole responsibility and liability of the user.

There has been a remarkable improvement in the safety performance of the South African mining industry since 2003. Many factors and initiatives have been responsible for this, but ultimately, they all find expression in action taken *by operational management* on mines. Ideally, this action should be voluntary and originate from on-mine considerations of how best to *provide and maintain a working environment that is safe and without risk to the health of employees*. Although legislative or other top down dictates may sometimes result in short-term improvements, if the action taken involves reluctant compliance, it is unlikely to stand the test of time.

It is in this context that the MOSH Leading Practice Adoption System has an important role to play, in that it is focused on achieving conditions that lead to voluntary and eager adoption of identified leading practice. Cutting edge techniques derived from behaviour science have been embedded in the detailed systematic approach of the adoption system to achieve this intrinsically sustainable outcome. In this sense, the MOSH Adoption System is a leading practice in its own right. Its potential contribution in addressing the lesser improvements achieved to date in the area of occupational health, as well as the substantial safety challenge that still remains, should be clear to those committed to achieving zero harm work places in the South African mining industry.

The primary purpose of this paper is to present the MOSH Leading Practice Adoption System in a readily understood format. A graphical style presentation with brief supporting notes is used for this purpose.

## *Motivating context for development of the System*

### *2013 Tripartite Industry Milestones*

**MINING INDUSTRY TRIPARTITE OCCUPATIONAL HEALTH AND SAFETY TARGETS AND MILESTONES SET IN 2003**

OCCUPATIONAL SAFETY
*Industry Target: Zero rate of fatalities and injuries*
*Milestones:*

- **In the Gold Sector**: By 2013 achieve safety performance levels equivalent to current international bench marks for underground metalliferous mines, at the least;
- **In the Platinum, Coal and Other Sectors**: By 2013 achieve constant and continuous improvement equivalent to current international benchmarks, at the least.

OCCUPATIONAL HEALTH
*Industry Target: Elimination of Silicosis*
*Milestones:*

- By December 2008, 95% of all exposure measurement results will be below the occupational exposure limit for respirable crystalline silica of 0.1 mg/m$^3$ (these results are individual readings and not average results)
- After December 2013, using present diagnostic techniques, no new cases of silicosis will occur amongst previously unexposed individuals (Previously unexposed individual = individuals unexposed prior to 2008, i.e. equivalent to a new person entering the industry at 2008).

*Industry Target: Elimination of Noise Induced Hearing Loss (NIHL)*
*(The present noise exposure limit specified in regulation is 85dB(A))*
*Milestones:*

- After December 2008, the hearing conservation programme implemented by industry must ensure that there is no deterioration in hearing greater than 10% amongst occupationally exposed individuals
- By December 2013, the total noise emitted by all equipment installed in any workplace must not exceed a sound pressure level of 110dB (A) at any location in that workplace (includes individual pieces of equipment).

**Notes**

1. The MOSH Leading Practice adoption system was developed to assist industry in its efforts to meet the Tripartite agreed Occupational Health and Safety Milestones.
2. The milestones were agreed to in 2003 and in 2005 the CEOs of the major mining companies committed their companies to achieve the 2013 milestones.
3. The MOSH Adoption system was developed in 2007 and piloted in 2008.

2 

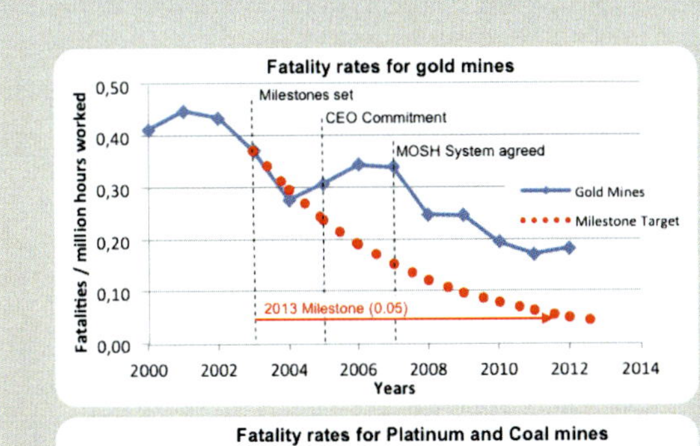


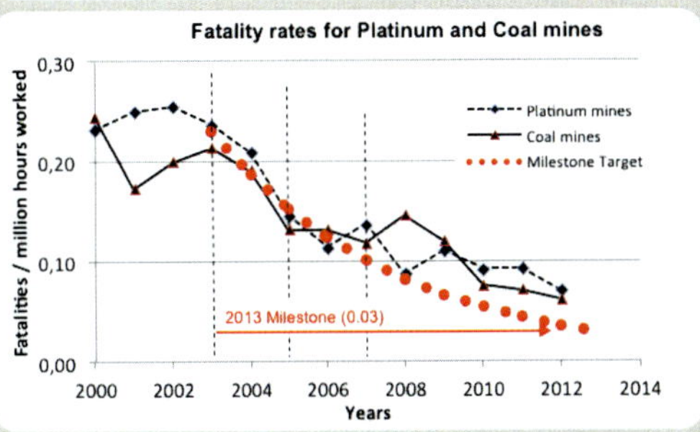


4. Industry has made considerable progress since 2003, but attainment of the milestones will require further sustained effort for the foreseeable future.

***Key features of the adoption system – industry ownership and behaviour modification***

***Industry ownership – schematic of the MOSH structures***

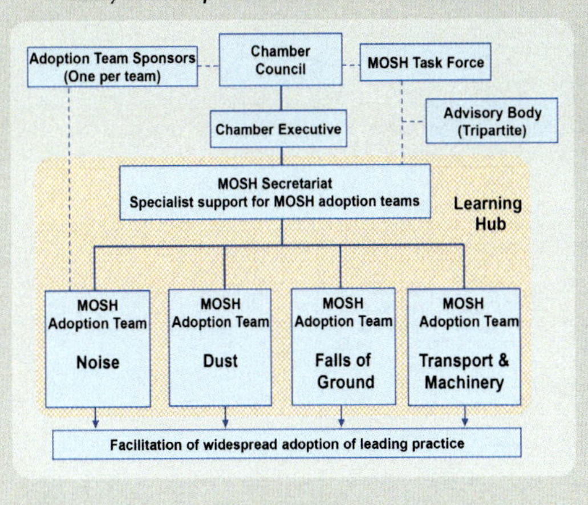


5. The Task Force is comprised of senior executives from mining companies. It decides on the major areas of risk to be addressed by the adoption system.

6. The MOSH Adoption Teams are led by experienced persons seconded to the Learning Hub by mining companies. The teams, comprised of representatives from the mining companies, are supported by a specialist secretariat.

7. Leading Practices having the greatest OHS improvement potential are selected at a workshop involving technical experts drawn from the mining companies and other organisations as appropriate.

*Schmatic of the behaviour change process*

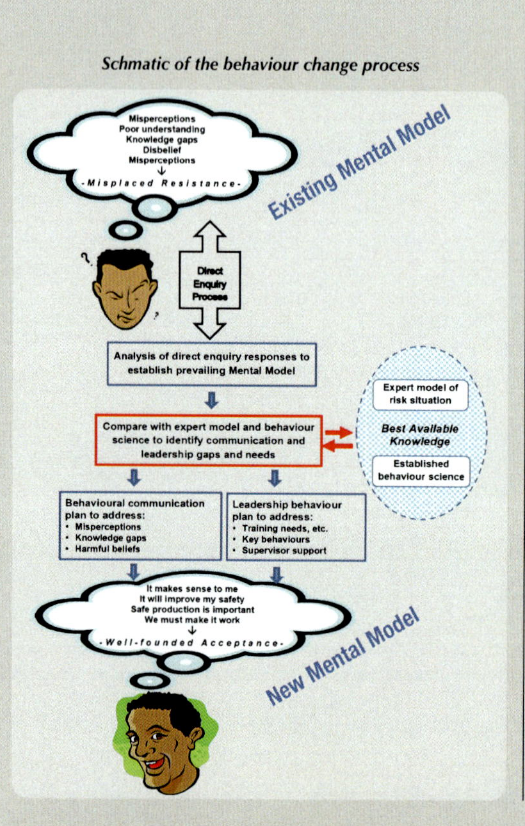


8. The adoption system recognises that a change in people's behaviour is fundamental to the adoption of any new practice or technology. Without this, adoption will not occur.
9. It also recognises that sustainable adoption requires prevailing mental models, which can act as barriers to adoption, to be identified and addressed through appropriate behavioural communication and leadership behaviour programmes.
10. Importantly, the required behavioural programmes are derived from a comparison of the prevailing mental model with scientifically established best available knowledge.
11. Industry ownership and behaviour modification are two key features of the MOSH Leading Practice Adoption System. Together they distinguish it from previous approaches at facilitating the transfer and adoption of new technology or practice.

## *Simple logic of the MOSH Adoption process*

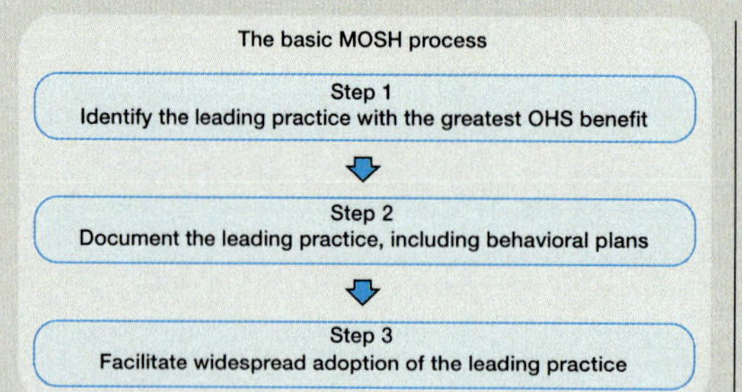


12. The essence of the MOSH Adoption Process is extremely simple. It is not new. What makes it special is the detail – full implementation of the detail is thus most important.

4 CM Mosh Leading Practice Adoption System – a leading practice in its own right S Malatji and JM Stewart

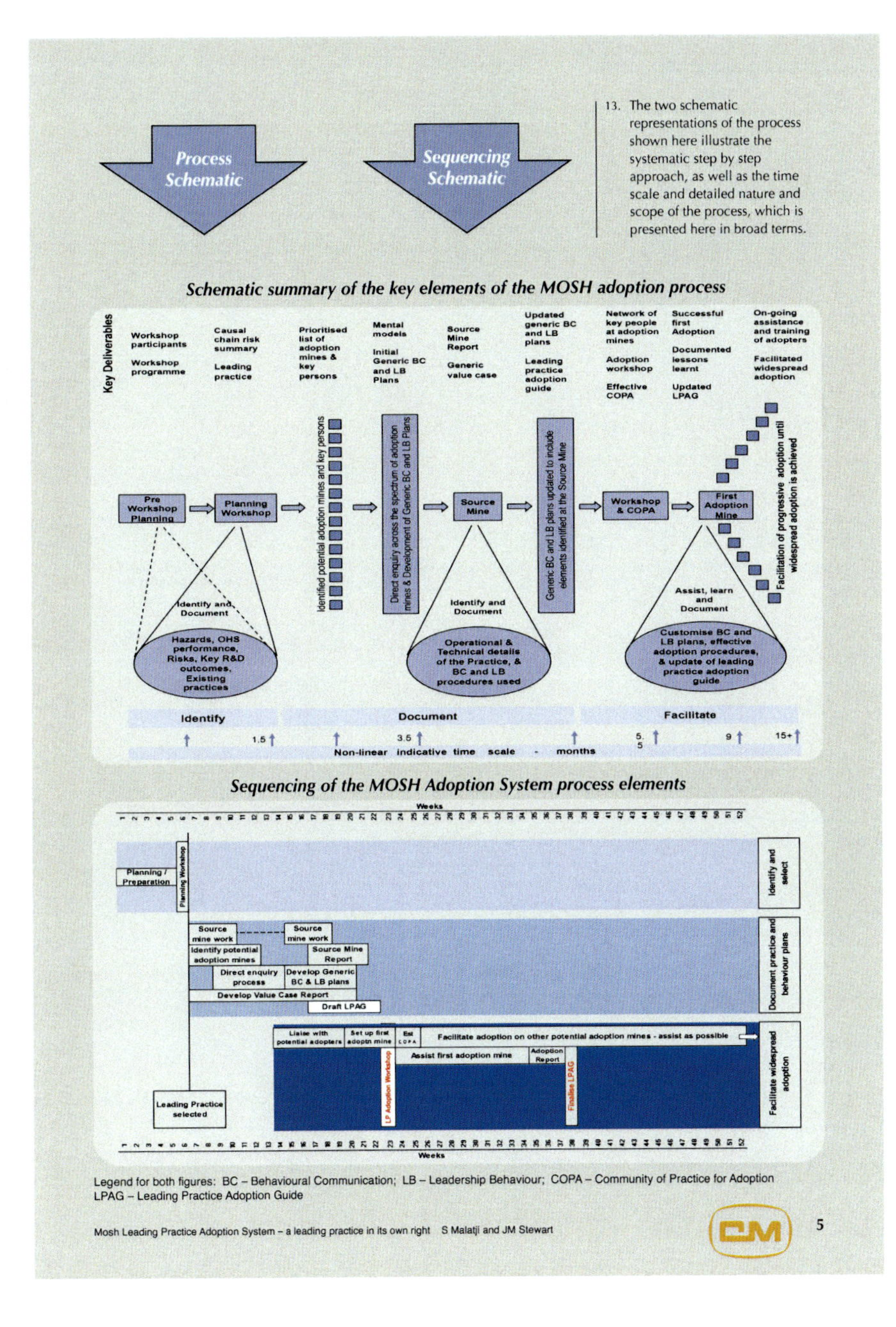
Process Schematic
Sequencing Schematic
13. The two schematic representations of the process shown here illustrate the systematic step by step approach, as well as the time scale and detailed nature and scope of the process, which is presented here in broad terms.
Schematic summary of the key elements of the MOSH adoption process
Key Deliverables
Workshop participants
Workshop programme
Causal chain risk summary
Leading practice
Prioritised list of adoption mines & key persons
Mental models
Initial Generic BC and LB Plans
Source Mine Report
Generic value case
Updated generic BC and LB plans
Leading practice adoption guide
Network of key people at adoption mines
Adoption workshop
Effective COPA
Successful first Adoption
Documented lessons learnt
Updated LPAG
On-going assistance and training of adopters
Facilitated widespread adoption
Pre Workshop Planning
Planning Workshop
Identified potential adoption mines and key persons
Direct enquiry across the spectrum of adoption mines & Development of Generic BC and LB Plans
Source Mine
Generic BC and LB plans updated to include elements identified at the Source Mine
Workshop & COPA
First Adoption Mine
Facilitation of progressive adoption until widespread adoption is achieved
Identify and Document
Hazards, OHS performance, Risks, Key R&D outcomes, Existing practices
Identify and Document
Operational & Technical details of the Practice, & BC and LB procedures used
Assist, learn and Document
Customise BC and LB plans, effective adoption procedures, & update of leading practice adoption guide
Identify
Document
Facilitate
1.5
3.5
5.5
9
15+
Non-linear indicative time scale - months
Sequencing of the MOSH Adoption System process elements
Weeks
Planning / Preparation
Planning Workshop
Identify and select
Source mine work
Source mine work
Identify potential adoption mines
Source Mine Report
Direct enquiry process
Develop Generic BC & LB plans
Develop Value Case Report
Draft LPAG
Document practice and behaviour plans
Liaise with potential adopters
Set up first adoptn mine
Est COPA
Facilitate adoption on other potential adoption mines - assist as possible
Assist first adoption mine
Adoption Report
LP Adoption Workshop
Finalise LPAG
Leading Practice selected
Facilitate widespread adoption
Weeks
Legend for both figures: BC – Behavioural Communication; LB – Leadership Behaviour; COPA – Community of Practice for Adoption
LPAG – Leading Practice Adoption Guide
Mosh Leading Practice Adoption System – a leading practice in its own right S Malatji and JM Stewart
CM
5

## *Ongoing provision of specialist support*

**Permanent full-time specialist secretariat to support Adoption Teams and Process**

- Adoption Specialist for each Adoption Team
- Behaviour Specialist
- Monitoring Specialist
- Administrative support

14. In order to ensure that the industry led MOSH Adoption Teams are fully effective in executing the MOSH adoption process they are supported by a dedicated full-time secretariat with the necessary specialist and administrative skills. *This is a very important enabling aspect of the adoption system.*

## *Establishment of MOSH Adoption Teams in key risk areas*

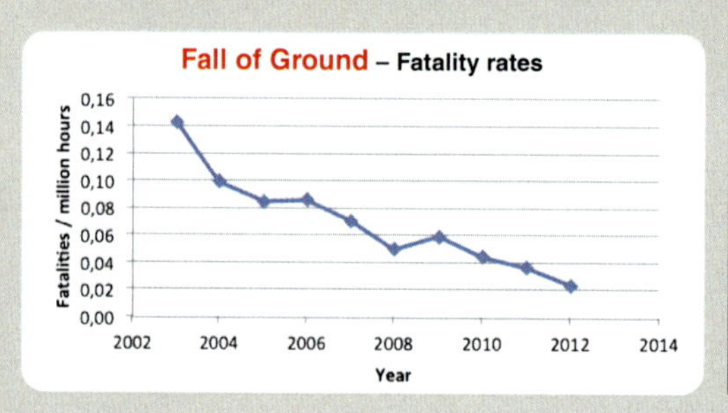


15. There has been a remarkable improvement in FOG fatality rates since 2002. It is no longer the stand-out leading cause of fatalities in SA mines.

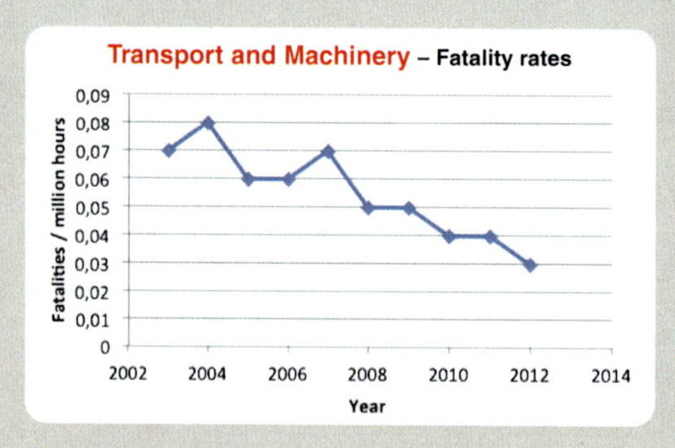


16. Transport and Machinery accidents as a cause of fatalities in SA mines is now comparable with those caused by falls of ground. With increasing mechanisation the importance of this risk area is likely to grow, and efforts in this area will need to be intensified.

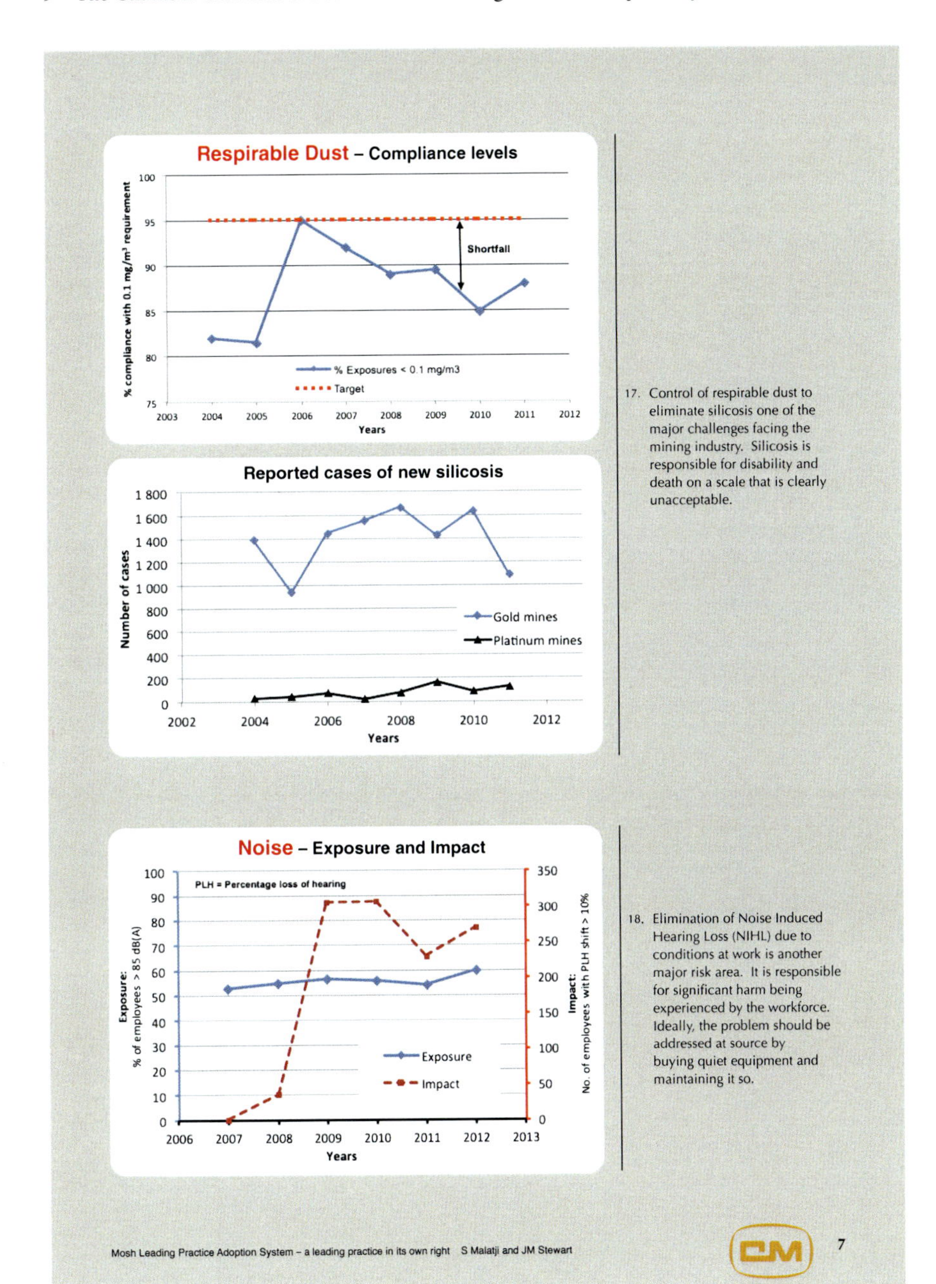
Respirable Dust – Compliance levels
% compliance with 0.1 mg/m³ requirement
Shortfall
% Exposures < 0.1 mg/m3
Target
Years
Reported cases of new silicosis
Number of cases
Gold mines
Platinum mines
Years
17. Control of respirable dust to eliminate silicosis one of the major challenges facing the mining industry. Silicosis is responsible for disability and death on a scale that is clearly unacceptable.
Noise – Exposure and Impact
PLH = Percentage loss of hearing
Exposure: % of employees > 85 dB(A)
Impact: No. of employees with PLH shift > 10%
Exposure
Impact
Years
18. Elimination of Noise Induced Hearing Loss (NIHL) due to conditions at work is another major risk area. It is responsible for significant harm being experienced by the workforce. Ideally, the problem should be addressed at source by buying quiet equipment and maintaining it so.
Mosh Leading Practice Adoption System – a leading practice in its own right S Malatji and JM Stewart
CM
7

## Developing an expert understanding of the risk situation

### *Basic Framework of a Generic Causal Chain Risk Summary Table*

| No | Nature of the hazard | No | Exposure to the hazard | No | Outcomes of exposure |
|---|---|---|---|---|---|
| **Part A – Description of the causal chain** | | | | | |
| | Description of the identified hazard/s (substances / equipment / events / etc.) | | Description of the different categories of exposed persons, the nature and duration of exposure, and any other details that characterise the exposure | | Description of all the ways in which workers might be affected and harmed by exposure to the risk, including biological or physiological effects. |
| | | | **Data Gaps** | | |
| | Identify any gaps that need to be investigated | | Identify any gaps that need to be investigated | | Identify any gaps that need to be investigated |
| **Part B - Current risk mitigation controls and strategies** | | | | | |
| | Identify and describe | | Identify and describe | | Identify and describe |
| | | | **Weaknesses** | | |
| | Identify and describe | | Identify and describe | | Identify and describe |
| **Part C – Possible improvements in risk mitigation controls and strategies** | | | | | |
| | Identify and describe | | Identify and describe | | Identify and describe |
| | | | **Possible new practices** | | |
| | Identify and describe | | Identify and describe | | Identify and describe |

19. The need for an expert understanding of the risk area being addressed is obvious. Without such an understanding the risk of addressing symptoms and not causes is real.

20. In the adoption system there is another equally important reason. Unless one has such an *expert* understanding, it is not possible to identify the knowledge gaps, misperceptions and mistaken beliefs of adopters and key stakeholders. It is these mental models that can act as barriers to adoption of a selected leading practice.

21. In the adoption system this expert understanding is captured in an expert model. The model may take the form of a causal chain risk summary table and / or an influence diagram.

### *Example expert model influence diagram - for the dust risk*

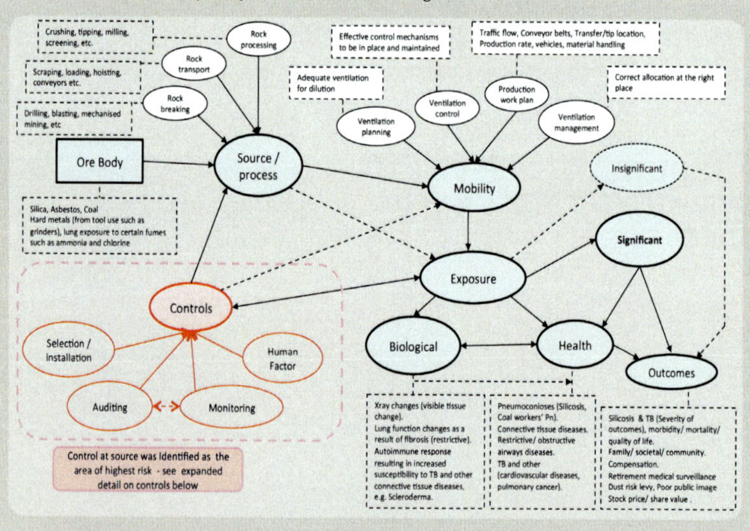


8 CM Mosh Leading Practice Adoption System – a leading practice in its own right S Malatji and JM Stewart

## Selection of the leading practice with greatest potential

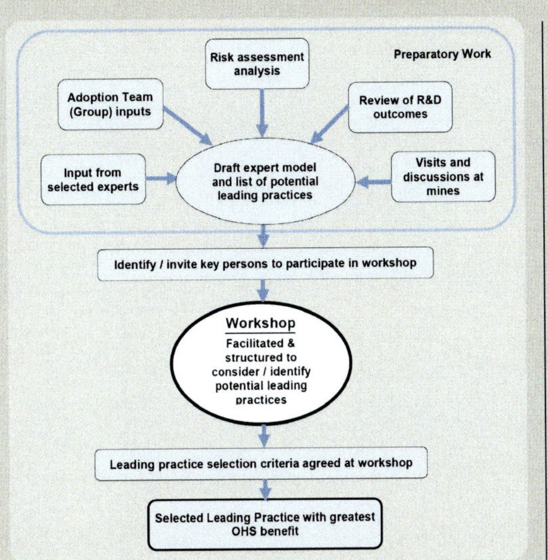


22. A special workshop involving the most experienced and knowledgeable people in industry is held to select the leading practice that has the greatest potential to improve OHS in the risk area in question.
23. The practice is selected using a rigorous procedure and relevant selection criteria agreed upon at the workshop.

## Identification of potential adoption mines and their key persons

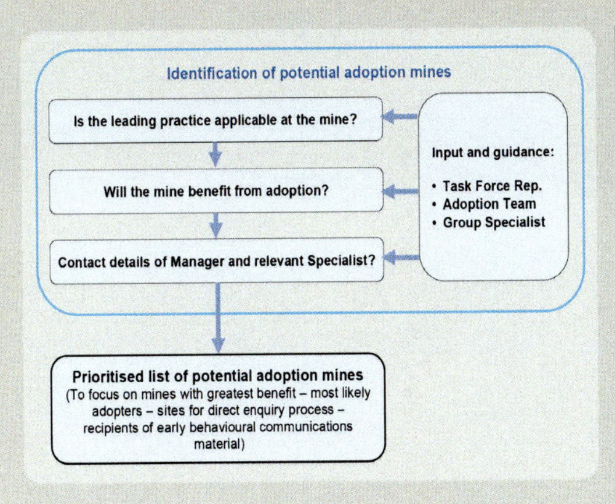


24. With input from the Task Force and mining company representatives of the MOSH Adoption Team, all potential adoption mines, and the key people at these mines are identified.
25. This enables:
    - The identification of key adopters, decision makers and other key stakeholders.
    - The selection of an appropriate group of persons to be interviewed in the direct enquiry process to establish the prevailing mental models.
26. A communication process to inform key persons of relevant findings and developments is then also possible.

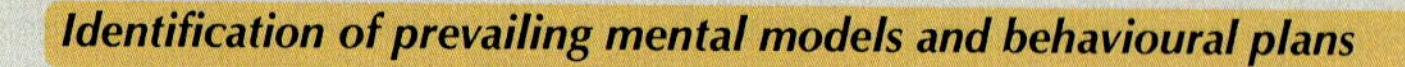

## Identification of prevailing mental models and behavioural plans

Identify key adopters and other stakeholders

Develop direct enquiry protocol (procedures / questions)

Select persons to be interviewed

Train interviewees and conduct enquiries

Analyse responses and establish mental models

Review mental model against expert model and behaviour science

Develop Behavioural Communication and Leadership Behaviour plans

27. The direct enquiry process recognises that all interviewees are experts about what they experience. It is thus a respectful process. It is used to identify the mental models of the key adopters, deciders and other stakeholders.
28. The carefully developed open ended questions of the direct enquiry process provide the interviewees with an opportunity to confidentially say exactly what they think.
29. The empirically determined mental models provide the basis for developing behavioural communication and leadership behaviour plans that are generically applicable to the complete spectrum of identified potential adoption mines.

## Documenting the leading practice at its Source mine

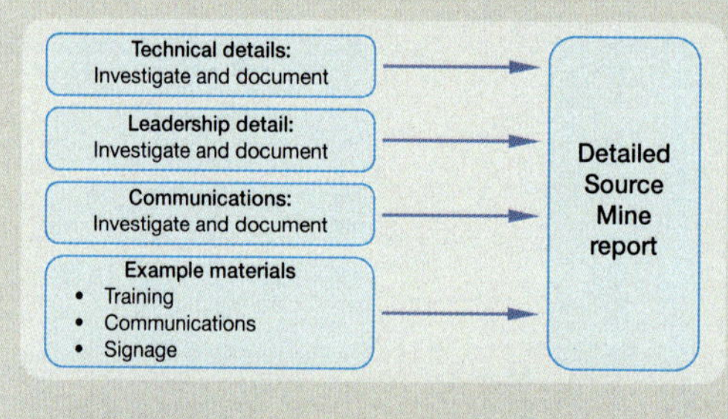


30. A detailed plan is developed for conducting the investigations at the source mine.
31. Investigation of the behavioural aspects is based on the insights gained in developing the generic behavioural plans.
32. The behavioural aspects observed at the source mine are used to update the generic behavioural plans.
33. The investigations seek to capture all the information needed to enable the practice to be replicated at another mine.

## Identifying and documenting the full value case

Financial Impacts
- Initial cost
- Operational cost
- Financial benefits
- Production
- Etc.

Strategic Impacts
- Health & Safety
- Relationships
- Image
- Investor support
- Etc.

The Generic Value Case

34. The value case is determined for each leading practice through a process of careful investigation.
35. The value case includes business case considerations but it also identifies the often more important strategic benefits associated with adoption of the leading practice.
36. Setting the goal of achieving zero harm at work is aspirational and admirable: adopting and appropriately acting on it requires high-level strategic decision making.

10 CM Mosh Leading Practice Adoption System – a leading practice in its own right S Malatji and JM Stewart

## Provision of a Leading Practice Adoption Guide to facilitate widespread adoption

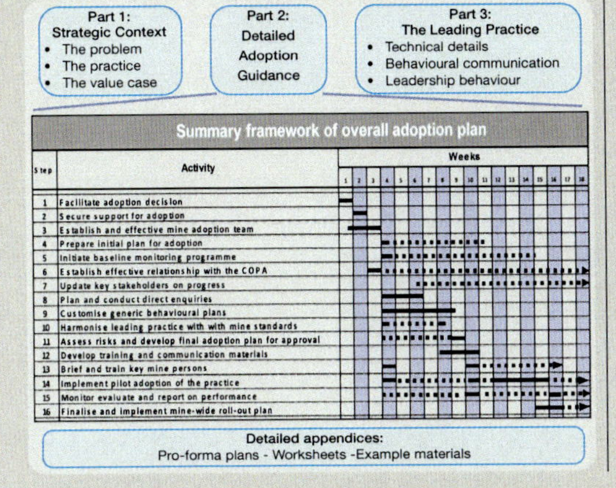


37. The Leading Practice Adoption Guide (LPAG) is a user-friendly document that contains all the information that a potential adoption mine needs to voluntarily decide whether or not to adopt the practice, and to then successfully adopt the practice should it decide to do so.

38. The Guide, particularly Part 3, makes it clear that the leading practice comprises of three equally important legs. All three need to be implemented for adoption to be complete and sustainable.

## Customisation of the leading practice at adoption mines

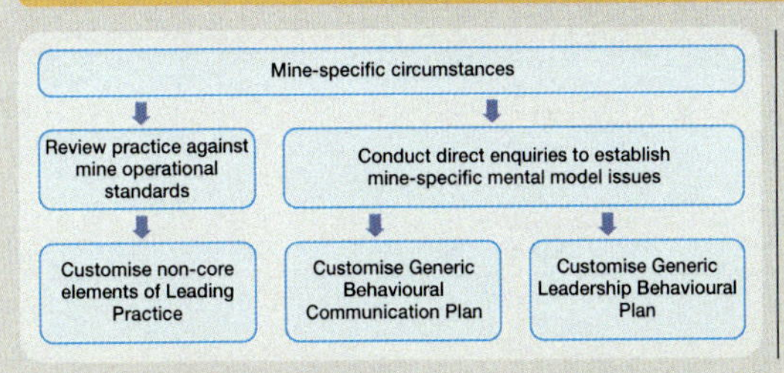


39. The adoption procedures outlined in the Leading Practice Adoption Guide include guidance on customising the leading practice to fit in with mine-specific circumstances.

## Initiation of widespread adoption process – the Leading Practice Adoption Workshop

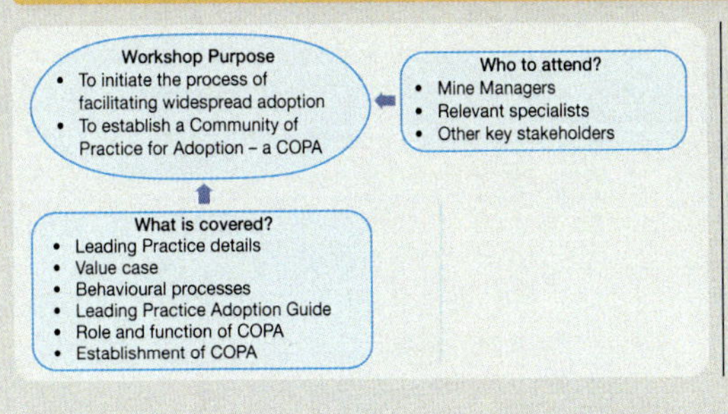


40. Key persons from all identified potential adoption mines are invited to attend the Leading Practice Adoption Workshop.

41. The workshop, which leads to the establishment of a Community of Practice for Adoption (a COPA) actively starts the process of facilitating industry wide adoption of the selected leading practice.

Mosh Leading Practice Adoption System – a leading practice in its own right S Malatji and JM Stewart

11

## Establishment of a Community of Practice for Adoption – for on-going facilitation

**Who?**
- Mine Managers (initially)
- Relevant mine specialists
- Mine Adoption Team Manager
- Mine Behavioural Overseer

**COPA**
Provide ongoing facilitation of the adoption process

**What?**
- Explain / use Leading Practice Adoption Guide
- Provide / arrange training
- Problem solving
- Provide / arrange assistance
- Share adoption experience
- Continuous improvement
- Mine visits as appropriate

**Modus operandi?**
- Regular meetings
- Jointly implement Leading Practice Adoption Guide
- Work directly with lead adopter mine
- Assist other adoption mines to extent possible

42. The COPA serves as a mechanism for mines to acquire guidance, assistance and specialist training to achieve successful adoption of the Leading Practice.
43. The COPA terminates its existence when its members feel that it is no longer needed.

## Conclusion – the process is continuous

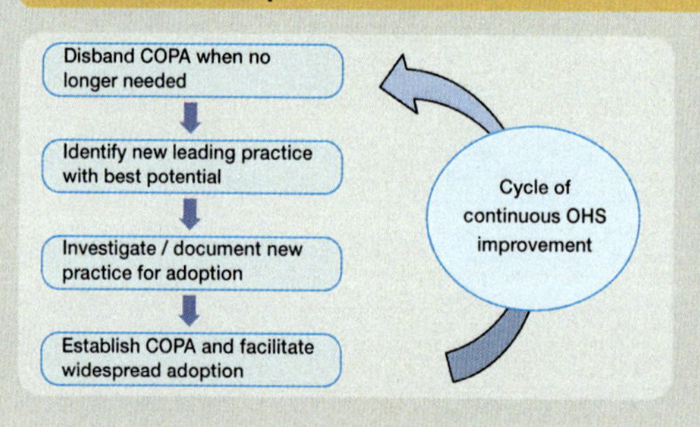


44. The process of continuous improvement never stops.

**Acknowledgement**
The role and contribution of the MOSH Adoption Teams, the Learning Hub Secretariat and Chamber member mining companies to the development of the MOSH Leading Practice Adoption System is gratefully acknowledged. So too is the contribution of Decision Partners LLC, Pittsburgh, USA, particularly in regard to development of the behaviour-based techniques that form a key element of the system. Much of the material presented in this paper is drawn from the Guidance Handbook on the MOSH Leading Practice Adoption System prepared by J M Stewart on behalf of the Chamber of Mines of South Africa.

***For further information and contact details, please see***
***www.mosh.co.za***

## Reference

Stewart, J. M. (2014). MOSH leading practice adoption system guidance handbook, revision 4.1. Chamber of Mines of South Africa.

# Chapter 10
# Conducting Effective Outreach with Community Stakeholders About Biosolids: A Customized Strategic Risk Communications Process™ Based on Mental Modeling

**Sara Eggers and Sarah Thorne**

## Introduction

When properly treated and managed, wastewater by-products, termed biosolids, can be applied to land as a beneficial soil amendment rather than being disposed of through landfills or other nonsustainable practices. The long-term viability and sustainability of biosolids land application requires community stakeholders' trust and support in the professionals who produce and apply biosolids in their communities. The Strategic Risk Communications Process™[1] (described in this chapter; Health Canada 2006), an application of the Mental Modeling approach, was customized and tested as part of a 2008–2011 research challenge supported by the U.S. Water Environment Research Foundation (WERF).[2] The project goal was to develop the process, methods, tools, and materials to enable biosolids managers to conduct effective outreach and dialogue with key stakeholders in the communities where they operate. The Process was applied and validated through two case studies conducted in collaboration with the City of Tulsa Public Works and the Virginia

[1] The Strategic Risk Communications Process™ was developed by Decision Partners Inc. for Health Canada and the Public Health Agency of Canada in 2006 and has been adapted for use by WERF and its members.

[2] This chapter is based on the reports developed in the course of this project, including A Strategic Risk Communications Process for Outreach and Dialogue on Biosolids Land Application © 2011 WERF and Conducting Effective Community Outreach and Dialogue on Biosolids Land Application: Primer for Biosolids Professionals © 2011 WERF and © 2011 Decision Partners, LLC. Permission granted by all copyright holders.

S. Eggers, Ph.D. (✉)
U.S. Food and Drug Administration, 8323 Haddon Drive, Tacoma Park, MD 20912, USA
e-mail: saralynneggers@gmail.com

S. Thorne, M.A.
Decision Partners, 1084 Queen Street West, #32B, Mississauga, ON, Canada L5H 4K4
e-mail: sthorne@decisionpartners.com



M.D. Wood et al., *Mental Modeling Approach*, Risk, Systems and Decisions,
DOI 10.1007/978-1-4939-6616-5_10

Biosolids Council. As part of the Process, mental models research was conducted with local Landowners who receive biosolids, Neighbors to biosolids land application sites and, in one case study, regional state Public Health Officials (PHOs). The research showed that these stakeholders were generally willing to support local biosolids programs, but they wanted their questions and concerns about odor, safety, and fairness to be appropriately addressed (North East Biosolids and Residuals Association 2007). Actionable outreach and dialogue plans and pretested communications materials were developed based on the research findings and the needs of the specific biosolids programs. The adapted Process and case study learnings were integrated into a comprehensive primer for biosolids managers to adopt the Process within their own programs. This work was intended to significantly improve biosolids professionals' ability to design and conduct effective outreach and dialogue with their community stakeholders.

## The Global Opportunity for Biosolids Professionals

Biosolids are the primarily organic solid product yielded by municipal wastewater treatment that has been processed to meet the land-application standards set forth by the U.S. Environmental Protection Agency (EPA; 40 CFR Part 503). In the United States, seven million dry tons of biosolids are produced annually by more than 16,000 municipal wastewater treatment facilities. Approximately 50 % of biosolids are recycled for land application on less than one percent of the nation's agricultural land.[3]

The production, management, and use of biosolids are a complex operation, supported and sustained by the decisions and actions of many different stakeholders. These include biosolids professionals, such as municipal utility staff, land appliers, and other contractors who manage production and safe use, as well as regulators who set standards and oversee their production and use. Importantly, it also includes stakeholders in the local communities where biosolids are produced and applied: farmers who benefit from this product, local officials who determine whether and how land application is acceptable in their community, and community members who give input on the acceptability of its use in their neighborhoods.

Stakeholders' judgment regarding the appropriate use of biosolids is fundamental to the future of biosolids operations across North America and in other countries (National Biosolids Partnership 2005; Pepper et al. 2008). By appropriate use, we mean that biosolids are produced and land applied in a way that maximizes the benefits of land application while ensuring minimal risk to people and the environment. However, biosolids managers and researchers have long identified significant challenges associated with individual, group, or community objections to biosolids land application (Beecher et al. 2004; Pepper et al. 2008).

[3] https://www.epa.gov/biosolids/frequently-asked-questions-about-biosolids.

In 2008, the Water Environment Research Foundation (WERF) and its members recognized the need to develop systematic and science-based guidance and tools to help biosolids professionals more effectively manage and communicate the benefits and risks associated with biosolids land application. WERF contracted researchers at Decision Partners to address this need. The goal of the WERF Research Challenge[4] was to provide biosolids professionals the means to significantly improve their communications with stakeholders on the benefits and risks of biosolids land application, and in so doing, enable stakeholders' judgments on the appropriate use of biosolids in their communities. The researchers met this goal by adapting their science-based Strategic Risk Communications Process™ (described in the next section) with supporting methods (including mental models research), tools, and materials. This Process and tools can be customized and used by biosolids professionals to meet their own needs.

This risk communications project was conducted as part of a broader initiative to develop a risk assessment tool for assessing and communicating the risks of pathogens in biosolids land application. The researchers worked very closely with their collaborators[5] to implement an integrated approach to risk assessment and risk communications. The collaborators also served as technical and regulatory expert advisors to the researchers.

In the course of this project, two overarching opportunities became very clear. The first was the opportunity to increase the beneficial use of biosolids through land application. The rate of biosolids land application has not increased over the past decade, in large part due to concerns about community acceptance and, in response, tighter regulations and requirements put in place by state and local governments over the production and application of biosolids (Wardell 2010). Although other favorable options are emerging (e.g., energy reclamation), land application is still considered the most economically viable and environmentally sustainable management option for most municipalities.

To achieve the first opportunity, a second opportunity was critical: to better engage local community members in a dialogue about this beneficial use in their communities. Biosolids professionals have historically used a "decide-and-defend" approach to communications. As noted by Beecher et al. (2004, 46), "the biosolids manager, in consultation with the biosolids user, regulatory staff, and agricultural advisors, makes the decisions.... Once the decisions are made, the public just needs to be educated, so they will accept the decisions." However, research (e.g., Vajjhala and Fischbeck 2006) has shown that such "decide-and-defend" communications strategies too often fail because: the issues are too complex; the trustworthiness of the communicator may be questioned; the communications do not provide informa-

[4] The WERF Challenge—Biosolids Risk Assessment and Risk Communications (SRSK2R08) was awarded to Decision Partners (Risk Communications) and Drexel University (Risk Assessment).

[5] This team was led by Dr. Patrick Gurian of Drexel University, and comprised approximately ten technical and regulatory academic experts from several universities. Their report, entitled Site Specific Risk Assessment Tools For Land Applied Biosolids is available from WERF (www.werf.org).

tion that stakeholders really want or need to make a well-informed decision; and those communications often come too late in the process, leaving stakeholders with the perception that they have no say in decisions.

More effective communication strategies strive to gain stakeholders' thoughtful judgment and ultimately their trust. Stakeholders' trust in the people managing risks can have a significant influence on whether a person accepts a perceived risk and the way it is being managed. People's judgment of trust involves many highly complex factors, including the perceived competence and trustworthiness of decision-makers and risk managers, the transparency of communications and actions, and stakeholders' perceptions on their ability to have meaningful input into the decisions that they believe can affect them (Slovic 2000). This is especially true in a biosolids management context, where different perceptions of risk can be expected between those who clearly benefit from a certain biosolids management option, and those who live, work, or play in areas that might be adversely affected by it.

## The Strategic Risk Communications Process

Strategic Risk Communications™, an application of the Mental Modeling approach, is defined as the purposeful process of skillful interaction supported by appropriate information. Its purpose is to enable well-informed decision making and action on risks. It includes all messages, actions, and interactions that can influence the judgments, decisions, and actions of both decision-makers and stakeholders. This Process, based on the Strategic Risk Communications Framework and Handbook (Thorne et al. 2006) originally developed for Health Canada and the Public Health Agency of Canada, has been adapted and applied to address a broad range of risk management and risk communications challenges and opportunities in public health, occupational health and safety, energy policy, and corporate product stewardship.

In this project, the Strategic Risk Communications Process™ was adapted to address the unique needs of biosolids professionals and the circumstances that define the communications challenges and opportunities in the communities in which they operate. The process (Fig. 10.1) comprises six integrated steps. The following summarizes its application for WERF:

- **Step 1: Define the Opportunity:** A Biosolids Communications Team was formed. The Team developed an *Opportunity Statement*—a clear statement of the focal community outreach and dialogue opportunity for the biosolids program. The Team also developed a plan for implementing the Process and (optional) guiding principles for community engagement and dialogue.
- **Step 2: Characterize the Situation:** The Team conducted a preliminary *technical assessment* of the biosolids land application program in the community (or communities), identifying the various related technical, social, regulatory, environmental, economic, and safety considerations. The Team also conducted a *pre-*

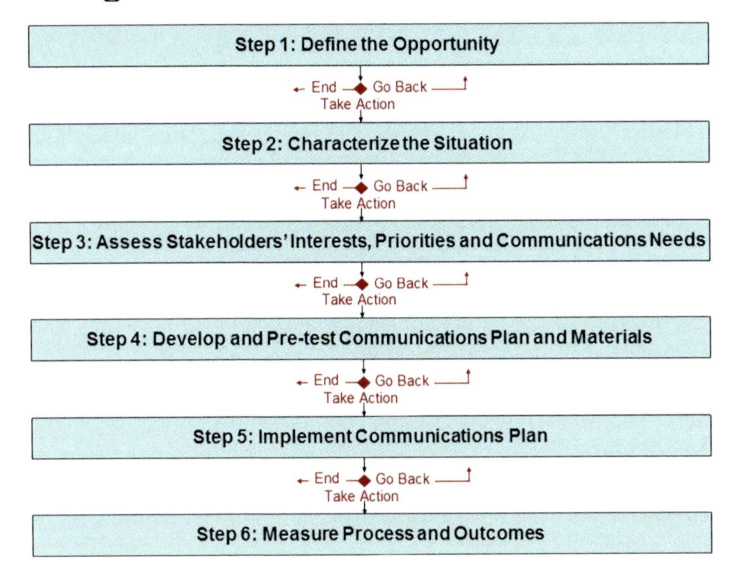


**Fig. 10.1** Strategic risk communications process for outreach and dialogue on biosolids land application

*liminary stakeholder assessment* to determine who the key stakeholders were and to capture the Team's current understanding of those stakeholders' perceptions and priorities related to biosolids land application in their community.

- **Step 3: Assess Stakeholders' Interests, Priorities, and Communications Needs:** Using a variety of formal or informal methods, including mental models research, the Team gained an in-depth and first-hand understanding of key stakeholders' perceptions, priorities, and communications needs. This new insight is analyzed against what was learned in Step 2 (*Characterize the Situation*). From that comparison, the Team defined the gaps and alignments that were to develop the Communications Plan in Step 4.
- **Step 4: Develop and Pretest Communications Plan and Materials:** Using the insight generated in the previous Steps, the Team develops a Communications Plan for community outreach and dialogue that addressed what the Team learned in Step 3 about people's perceptions, priorities, and communications needs.
- **Step 5: Implement Communications Plan:** The Team implemented the Plan developed in Step 4, monitoring, adjusting, and refining the Plan and materials as required based on what they learned through their outreach efforts about stakeholders' interests and priorities.
- **Step 6: Measure Process and Outcomes:** The Team measured the effectiveness of their outreach and dialogue efforts and determined the degree to which they achieved the Opportunity defined in Step 1. They also assessed the effectiveness of the Process and made recommendations for future Team efforts as part of their commitment to continuous learning and continuous improvement.

This customized Strategic Risk Communications Process™ can be tailored to address a wide range of challenges, opportunities, and situations. It can be adapted to fit the unique needs of a community's biosolids program or to provide an approach for an organization that conducts land application in many communities. It involves an iterative process with integrated steps, scalable to provide maximum value with the available time and resources. Finally, the Process supports continuous learning and continuous improvement. Because it is a systematic process, it can be applied repeatedly; building knowledge upon knowledge; and tracking how the perceptions, priorities, and communications needs of stakeholders change over time.

The key goal of the Process is to enable communication between biosolids professionals and their stakeholders. To be effective, these communications must be based on an in-depth understanding of stakeholder interests, priorities, and information needs related to those topics. Thus, the Strategic Risk Communications Process™ calls for science-based and purposeful research with both experts and stakeholders. The Mental Modeling approach is ideally suited to generate needed insight in both the expert and stakeholder domains because it provides systematic procedures for understanding what people (experts and stakeholders alike) know and need to know, and for empirically evaluating how well communications address those needs (Morgan et al. 2002).

## Applying the Strategic Risk Communications Process™: Two Case Studies

The customized Strategic Risk Communications Process™, supported by mental models research, was applied through two case studies. [6] The researchers first worked with a team from the City of Tulsa Public Works (Tulsa Team). The researchers then worked with a team comprising several members of the Virginia Biosolids Council (VA Team). Key objectives of the case studies were to: a) customize the Strategic Risk Communication Process™ to address the specific challenges and opportunities facing these two biosolids programs; and b) to inform the development of prototypical strategies that could be adapted by other biosolids professionals to serve their own needs. One goal of this project was to tailor the scope of the comprehensive mental models research to fit the individual needs and resources of biosolids land application programs, so that these programs could derive greatest value from the Process, within their time and resource constraints.

The rest of the chapter summarizes how the Process was applied in the case studies, emphasizing how mental models research supported the application of the Process. Key findings and products of the case studies and broader lessons for the biosolids industry are highlighted.

[6] Steps 1 to 4 were completed as part of the case study applications. The researchers offered the Case Study Teams specific recommendations and draft plans for completing Steps 5 and 6 of the Process on their own.

## Step 1: Define the Opportunity

The initial step in the Process is to form a Communications Team and define the biosolids outreach and dialogue opportunity. This is vital to ensuring that the team members have a clear statement and shared understanding of the common goal, how that goal will be achieved, and how success will be measured.

A Biosolids Team was formed in each case. In Tulsa, the Team comprised the lead manager, operations manager, and administrative manager from the City of Tulsa Public Works, as well as a research project advisor from WERF. In Virginia, the Team comprised seven representatives from the Virginia Biosolids Council member organizations, which included large and small community wastewater treatment facilities, biosolids management companies, and biosolids consulting services.

Through facilitated discussions, the Teams developed opportunity statements similar to the sample shown as follows. In both cases, the opportunities focused on gaining the local stakeholders' trust and confidence in the value that their programs provide, through community-focused outreach and dialogue.

## Sample Opportunity Statement

*CITY PUBLIC WORKS BIOSOLIDS PROGRAM OPPORTUNITY TO DEVELOP A COMMUNITY ENGAGEMENT DIALOGUE PROCESS*

*The success of the City Public Works Biosolids Program requires the trust and support of the people in the communities where our biosolids are land applied. The opportunity for the City Public Works Biosolids Team is to enable our community stakeholders to make well-informed judgments and decisions in support of our biosolids land application program by actively engaging them in meaningful dialogue supported by relevant information.*

*Over the next 6 months, we will apply the Strategic Risk Communications Process in order to identify the values, priorities, interests and informational needs of key community stakeholders regarding biosolids land application in the community. We will use the knowledge we gain to design and implement a stakeholder engagement process specifically focused on engaging in dialogue with our community stakeholders.*

*We will measure the success of our efforts by systematically requesting and assessing feedback from community stakeholders to:*

(a) *Determine the degree to which they feel that they had the opportunity to engage with our Team in a meaningful way; and*

*Measure the degree to which our stakeholders' confidence in our Team and our partners, our operations, our products and our services has increased.*

## Step 2: Characterize the Situation

With the opportunity identified, the next step in the Process was to gather and integrate the knowledge and hypotheses of the biosolids experts. This step was important to understanding the necessary context of the biosolids land application program in the community(ies), defining the various related technical, social, regulatory, environmental, economic, and safety considerations.

### *Technical Assessment, Using Expert Modeling*

The technical assessment was based on comprehensive literature review and input from the expert researchers and advisors who took part in the Research Challenge. In addition, each Case Study Team was facilitated through in-person group discussion on the key technical, regulatory, and other issues related to biosolids land application in their communities. Team members also provided summary documents describing these key issues in their local communities.

A comprehensive *Expert Model of the Influences on Integrated Risk Assessment and Risk Communication for Biosolids Land Application* was developed based on this input. This Expert Model provided a "systems" view of biosolids land application management, identifying the key stakeholders, processes, and potential outcomes in that system. The full Model included the entire range of influences on the management of biosolids, from production through its application, including the influence of community acceptance on the viability of land application within the community. Figure 10.2 is an excerpt of the Expert Model, focusing on specific impacts of biosolids land application on people (including landowners who apply biosolids and other people who are exposed to biosolids) and the environment.

The Expert Model was reviewed by the experts on the Research Team and was presented to biosolids professionals at a research forum of WERF members. The expert feedback indicated that the Model provided a comprehensive picture of the entire system of biosolids land application. Some also believed that the Expert Model could also be a useful tool for assessing and communicating opportunities to improve biosolids management among biosolids professionals. For this project, however, the Expert Model served the analytical framework for the design and analysis subsequent mental models research with community members.

### *Preliminary Stakeholder Analysis*

A key task in the Process is to identify stakeholders—individuals or groups who can affect, be affected by, or perceive themselves to be affected by decisions related to biosolids land application. An understanding of the breadth of stakeholders—and

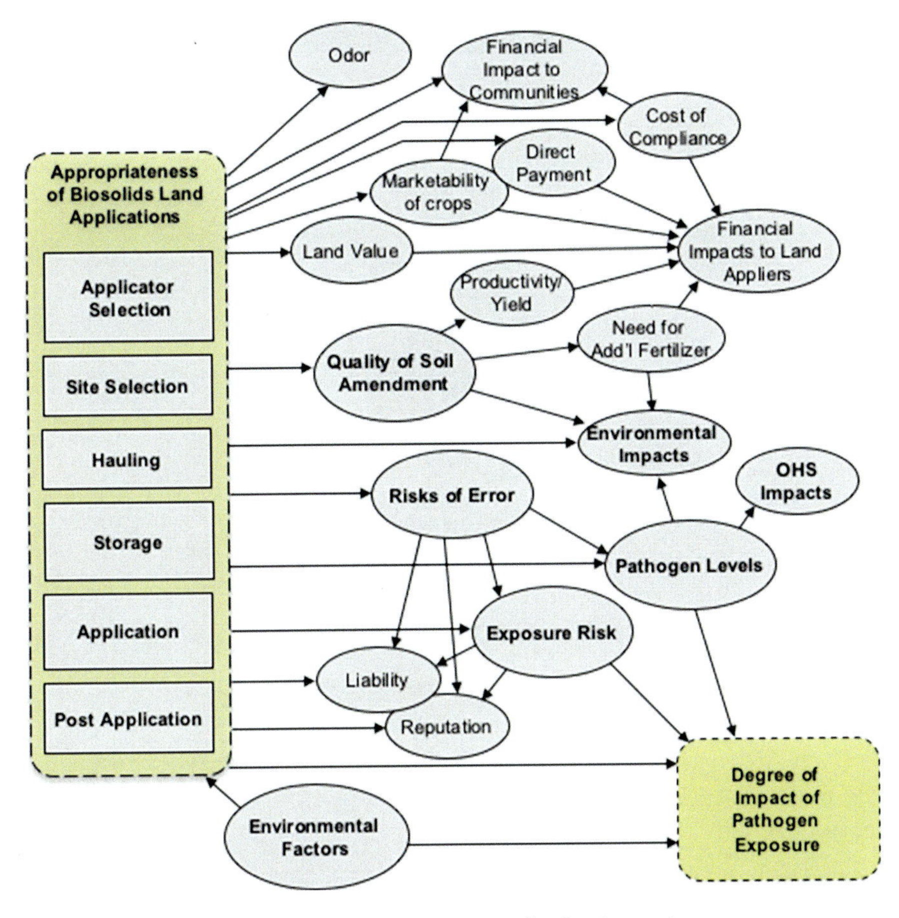


**Fig. 10.2** Expert model of impacts of biosolids land application (excerpt)

their role in decision making—is important to planning effective outreach and dialogue. A Stakeholder Map is an effective tool for identifying all of the stakeholders of biosolids land application. Figure 10.3 shows a sample Stakeholder Map of Biosolids Land Application which was developed based on the insight generated through expert model development.

At the center of the Stakeholder Map is the opportunity at hand; in this case, a sustainable biosolids land application program. The identified stakeholders—groups or individuals—have a relationship with a biosolids program that is more or less direct, depending on their proximity to the center. Stakeholders in the rings closest to the center are likely to have a direct "stake" in the management practices, decisions, and impacts related to biosolids land application.

With the universe of stakeholders to the biosolids land application program identified, the next task was to determine the key or high priority stakeholders for outreach and dialogue those stakeholders whose judgments and support are most critical to the

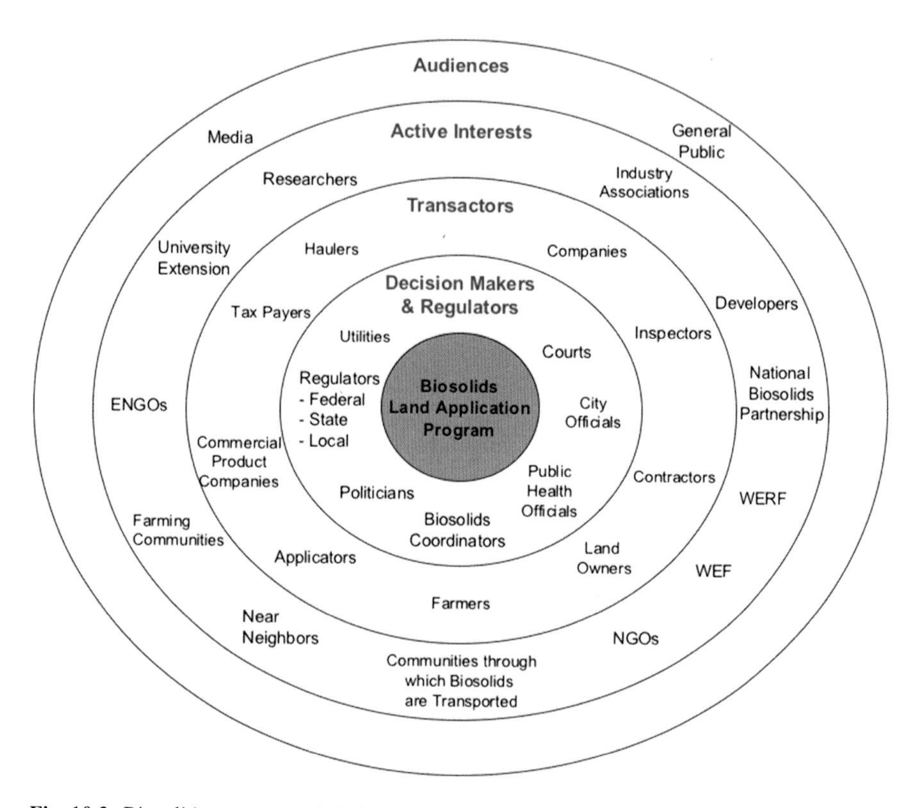


**Fig. 10.3** Biosolids program stakeholder map

success of the viability of the biosolids land application program in the local community, and whose priorities, values are communications needs might be most effectively addressed through community dialogue and outreach. Table 10.1 summarizes the key stakeholders identified by each Case Study Team. Near Neighbors (people who lived a few miles from a biosolids land application) were identified as key stakeholders by both Teams, because they may see themselves at being most affected by biosolids land application even though they do not receive any direct benefit. Members of the VA Team had also directly experienced neighbor complaints in the past.

The final task in conducting the preliminary stakeholder assessment was to develop hypotheses of the priority stakeholders' interests, priorities, and communications needs related to biosolids land application. A Stakeholder Hypothesis is the Team's "best guess" of these and it informs the preparation for research conducted with these priority stakeholders (Step 3 of the Process). In each case study, detailed hypotheses were developed for each priority stakeholder through facilitated discussions with the Team, drawing on their experiences in their community. These hypotheses were also compared against the results of the mental models research to identify gaps in the Team's understanding of their stakeholders' perceptions and priorities.

**Table 10.1** Identified key community stakeholders for sustainable biosolids land application programs

| Stakeholder | Type of stakeholder | Team's rationale for priority designation |
|---|---|---|
| Landowners (Tulsa Team) | **Transactors:** Receives free product as part of the program. | We see an opportunity for Landowners to become key partners in communication with the surrounding neighbors. We believe that they are well respected in this rural community and know the Near Neighbors well and are on friendly terms with them |
| Near Neighbors (Tulsa and VA Teams) | **Active Interests:** Do not receive any direct benefit from the program, but may perceive themselves to be affected by the local biosolids land application | We believe Near Neighbors are most affected by the local land application because they live closest to the sites and can be affected by the increased traffic and sometimes odor. There have been recent inquiries to the City Public Works office from rural citizens voicing their concerns about potential health impacts which they believe are caused by the odor. Some have also expressed concern about run-off of biosolids into the river |
| Virginia Department of Health (VDH) Officials (VA Team) | **Active Interests:** May take an active interest in issues regarding the safety of biosolids | VDH Officials have expressed interest in learning more about biosolids land application. We believe that they are well respected in this community and may be a direct source of information for community members. They may also have a strong influence on the local community leaders. Some VDH Officials have been vocal in their concerns about biosolids safety |

## Step 3: Assess Stakeholders' Interests, Priorities, and Communications Needs, Through Mental Models Research

With a full expert characterization of the situation, both the technical and social aspects, we proceeded to Step 3 of the Process: conducting purposeful and empirical research with target stakeholders to identify their interests, priorities, and information needs. With this insight, the Team was able to determine where its (as biosolids professionals) interests and priorities were aligned with those of its stakeholders and where there were gaps. Reinforcing the alignments and closing the gaps would form the basis for the Team's risk communications plans and materials in Step 4.

Empirical research, following the Mental Models approach, was conducted in both case studies. The objectives of this research were to: (a) elicit the values, priorities, interests, and informational needs related to biosolids land application of these key stakeholders in two geographic areas—Oklahoma and Virginia; (b) support development of risk communications tools and core messages that could be adapted by biosolids professionals; and (c) use the results of the case studies to develop prototypical communications materials.

## *Sample Development and Recruitment*

With well-defined cohorts and use of robust sampling development and recruitment methods, mental models research yields rich data from a small sample of individuals, typically on the order of 20–30 people (Morgan et al. 2002). Samples of this size do not allow for precise estimates of the prevalence of beliefs or attitudes held by people in the specific stakeholder population, nor can they be generalized to more general populations (e.g., U.S. consumers). However, with samples of this size we can be fairly confident that the research will yield discovery of the range of beliefs that are important for consideration in the development of communication strategies for the relevant stakeholder group.

The research samples were generated using two methods. In Tulsa, a list of current and past Landowners who receive land-applied biosolids was provided by the Case Study Team. A list of people who have expressed interest in biosolids land application to the City of Tulsa Public Works Department (referred to as potential landowners) was also provided. For the VA research with VDH Officials, a list of regional Department of Health Officials was provided by the Case Study Team. Participants were chosen at random and recruited to participate until the target sample sizes were met.

To develop the Near Neighbor samples, the Case Study Teams provided the location's address or intersections, or geographic coordinates of current or recent land application sites. The researchers identified streets within a specified distance from the application site, in any direction, using GoogleMaps. Yellowpages.com was used to identify residents with active landline phone numbers along those streets. Researchers chose randomly from the participant list, ensuring that there were no more than two Neighbors recruited who resided adjacent to a single application site. Participants were recruited to participate until the targets sample sizes were met.

## *Protocol Outline*

A semistructured interview protocol was developed based on the insight generated from the Technical Analysis (Expert Model) and the Stakeholder Hypotheses. The protocol aligned with the Expert Model as its analytical base. Key topics targeted stakeholders' perceptions and thoughts on:

- Biosolids in general, including, why and how biosolids may be land applied;
- Benefits and risks associated with biosolids land application and their impacts to land, water, animals, and people;
- The decision-making process for biosolids land application, including the roles and responsibilities of those who are involved; and
- Communications related to biosolids land application, including most important information in communities where biosolids are applied, and the most effective ways for biosolids managers to communicate about biosolids.

Two pretest interviews were conducted to ensure the protocol questions were interpreted as intended. Insight from the pretests was used to refine the protocol and prepare the analysis. The protocol tailored to meet the specific context of each Case Study Team.

## *Conducting Interviews*

Interviews were conducted by trained research interviewers oriented to this project. Potential participants were contacted by telephone and invited to participate in a telephone research interview. Potential participants were notified that they would be offered a $30 honorarium for participating and that their comments would be confidential to the research team. Interviews were scheduled with participants who agreed, at the participant's convenience.

Interviewers followed the interview protocol, asking specific prompts as required. Interviewees were encouraged to respond to questions as they understood them, as well as raise additional topics as they came to mind and to elaborate on their perspectives. Interviews were audio-recorded with the interviewee's permission, and near-verbatim transcriptions developed. To preserve confidentiality, names and other potentially identifying information were not included in the transcripts.

## *Coding and Analysis*

The Expert Model served as the analytical framework for coding and analysis. Interviewees' responses were parsed into discrete data segments. Each segment was assigned one or more codes that best aligned with specific expert model nodes. New codes were generated in cases where it was determined that model nodes did not sufficiently capture the theme as it was expressed by the interview response. Descriptive keywords were also assigned to capture finer details related to a particular node. Coded segments were considered "spontaneous" mentions if the interviewee's response was generated without prompting him/her to think of a particular topic. Segments were coded without consideration of the accuracy of the interviewee's statement. Initially, several interviews were coded independently by two individuals. Differences were resolved by discussion. Once the coding procedure seemed stable, a single individual coded subsequent interviews.

Coded interviews were transferred into a database where more detailed analyses were conducted on segments by interview question, code, keywords, or in combination in order to subjectively assess the results captured within each node and to identify other emerging themes. The coded results were organized into mental models diagrams representing the aggregate results from the interviews conducted in each cohort, in a format aligned with the expert model. Separate analyses were conducted for each of the research cohorts in order to identify potential variations in the perceptions and priorities between Landowners and Near Neighbors (and in VA, VDH Officials). For example, Figs. 10.4 and 10.5 shows mental models diagrams

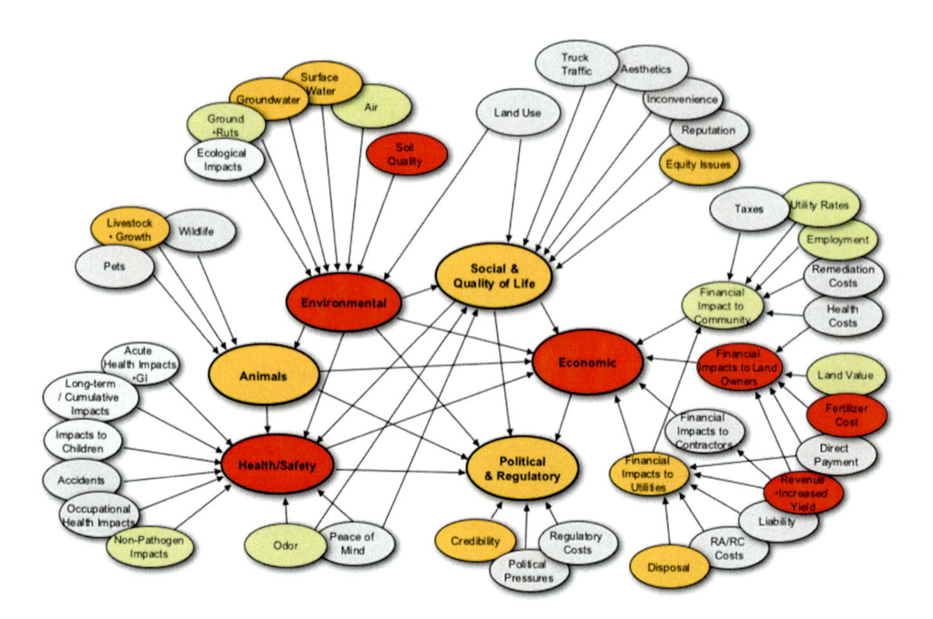


**Fig. 10.4** Mental models diagrams representing Tulsa Landowners' perceptions of the impacts from biosolids land application. *Red* nodes are themes identified as the likely strongest influence on interviewees' judgments, followed by *orange* and then light *yellow*. *Gray* nodes indicate themes not raised by interviewees in the specific research cohort

representing Tulsa Landowners' (10.4) and Neighbors' (10.5) perceptions of the impacts from biosolids land application. As shown in Fig. 10.5, mental model nodes were assigned a color (red, orange, light yellow) to indicate the likely strength of influence of the particular theme on interviewees' judgments overall. This qualitative assessment of the weighting of the themes was based on the frequency of interviewees' statements, coupled with the intensity of their statements. Also gray nodes indicate themes not raised by any interviewees in a specific cohort.

Additionally, the results were compared against the stakeholder hypotheses generated by each Case Study Team in order to assess what these stakeholders know that is correct, what they do not know or misunderstand that is consequential, what they want to know and who they trust, and what communications processes they trust. The gaps and alignments were then used to develop prototypical communications plans, messages, and materials.

## *Sample Characteristics*

Mental models interviews were conducted with: 12 Landowners and 12 Near Neighbors in Tulsa, Oklahoma; 24 Neighbors in three Virginia communities (Fauquier County, Tidewater area, and Lunenburg); and 6 VDH Officials (Table 10.2).

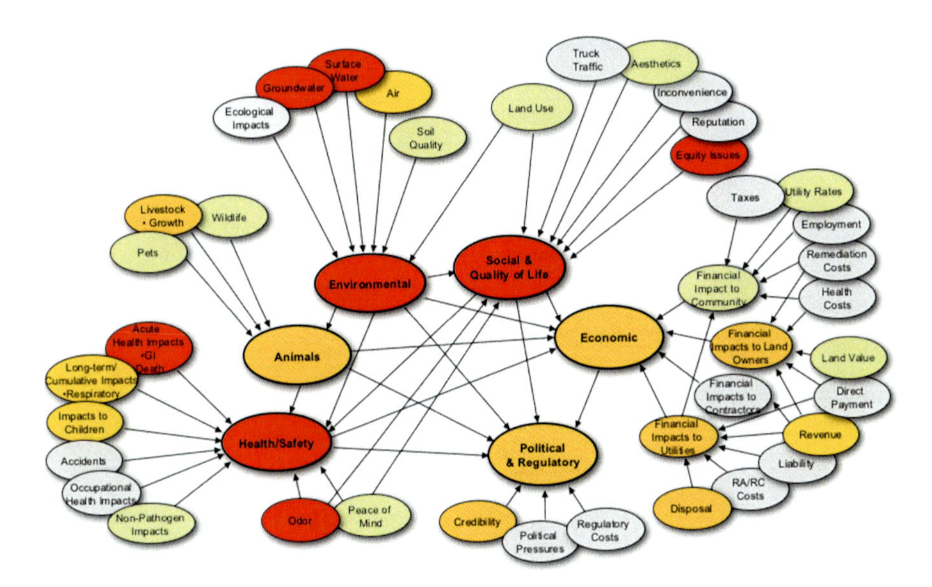


**Fig. 10.5** Mental models diagrams representing Neighbors' perceptions of the impacts from biosolids land application. *Red* nodes are themes identified as the likely strongest influence on interviewees' judgments, followed by *orange* and then light *yellow*. *Gray* nodes indicate themes not raised by interviewees in the specific research cohort

Research participants were generally very interested and willing to provide their thoughts on biosolids use within their community. They provided rich insight, which both Case Study Teams found very useful (and in cases surprising), supplementing their prior direct experiences with these stakeholders. Interviewees also provided in-depth insight as well as additional context to prior research on public perceptions of biosolids land application (e.g., Beecher et al. 2004; Krogmann et al. 2001; Robinson and Robinson 2006).

## *Highlights from the Mental Models Research*

The remainder of this section provides highlights from this research. (Note: These qualitative, small sample, and locally focused research results should not be interpreted as being representative of the U.S. population. Percentages are used only to indicate the proportion of the research sample that responded in a particular way.)

Most Tulsa Landowners and Near Neighbors demonstrated general familiarity with biosolids and biosolids land application. On the whole, interviewees in VA expressed less awareness of biosolids. More broadly, interviewees, particularly Near Neighbors, generally did not appear to have an in-depth understanding of water and wastewater systems. Some specifically commented that they were "not a part" of the wastewater system. When talking about biosolids, interviewees, particularly those with less familiarity, often drew on their perceptions of other envi-

**Table 10.2** Select interview characteristics

| | | Tulsa | VA |
|---|---|---|---|
| Near Neighbors | Attempted contact | 149 | 321 |
| | Accepted | 12 (8 %) | 24 (7 %) |
| | Declined | 68 (45 %) | 95 (29 %) |
| | Average interview length | 33 min | 41 min |
| | N (%) female | 3 (25 %) | 2 (17 %) |
| Landowners | Attempted contact | 30 | NA |
| | Accepted | 12 (40 %) | |
| | Declined | 12 (40 %) | |
| | Average interview length | 55 min | |
| | N (%) female | 13 (54 %) | |
| VDH Officials | Attempted contact | NA | 18 |
| | Accepted | | 6 (33 %) |
| | Declined | | 6 (33 %) |
| | Average interview length | | 45 min |
| | N (%) female | | 1 (17 %) |

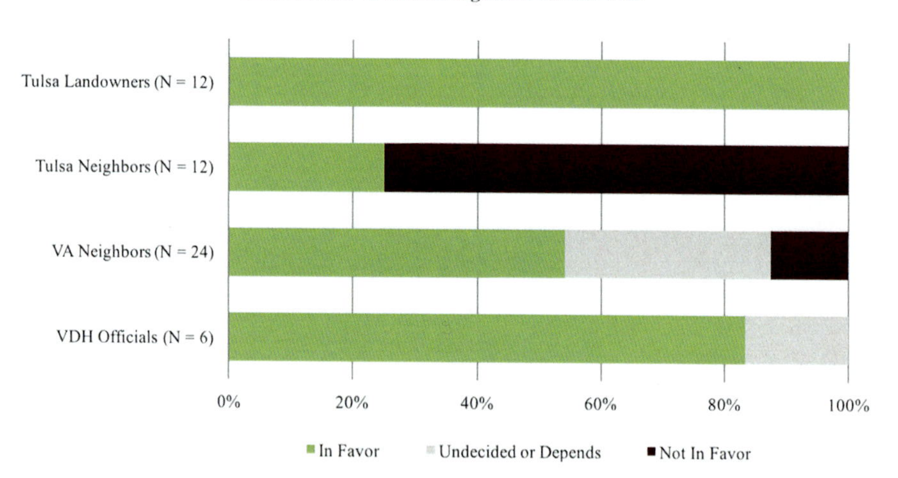


**Fig. 10.6** Interviewees' overall judgments on biosolids

ronmental and health issues—and how these issues were managed by the people in charge. Examples included the Gulf oil spill, PCB contamination in the Chesapeake Bay, animal manure, and commercial fertilizers.

Interviewees thoughtfully considered the perceived benefits, risks, costs, and trade-offs of biosolids land application in their communities. Near the end of the interview, interviewees were asked if they were generally in favor of biosolids land application in their county, generally not in favor of biosolids land application in their county, or undecided (Fig. 10.6). Tulsa Landowners were all "in favor" of biosolids land application. Tulsa Near Neighbors were generally "not in favor," citing the odor, health concerns, quality of life impacts, and uncertainty.

VA Neighbors were more varied in their judgments. About 50 % of the VA Neighbors were "in favor," citing the value to the agricultural community, responsible management of a societal problem, and confidence in biosolids management. About 10 % were "not in favor," citing the unacceptable uncertainty about health and environmental risks. The remainder (about 30 %) said they were "undecided" or "in favor with conditions." These interviewees wanted more assurance about the safety and long-term health and environmental impacts. Most VDH Officials commented that they were "in favor," generally referred to biosolids as the "best of both worlds."

Several common factors emerged as potentially having influence (positively or negatively) on interviewees' judgments on the acceptability of biosolids land application in their counties. These factors are described as follows, presented in an approximate order of "strength" of influence, as determined by a qualitative summary analysis across all cohorts:

***Individual's awareness and experience with biosolids.*** Those interviewees, particularly Tulsa Landowners and VDH Officials, who appeared to be more familiar with biosolids land application tended to say that they were "in favor." Some Neighbors said they felt "comfortable" with biosolids, because they grew up in an agricultural environment.

***Confidence in the safety of biosolids.*** Interviewees who were in favor of biosolids tended to express confidence that biosolids are safe. Tulsa landowners cited their trust in the City of Tulsa and its regulations, as well as the "naturalness" of biosolids versus "man-made" products. VDH Officials generally expressed confidence that "when all the rules are followed, [biosolids land application] appears to be a safe procedure."

Near Neighbors in Tulsa and VA who were "not in favor" or "undecided" expressed much less confidence in the safety of biosolids land application, citing a lack of confidence in the decision-makers and the regulations, and the uncertainty and "newness" of biosolids. These interviewees also talked about "not seeing enough data" or "being ignorant" of the risks to be able to conclude that biosolids are safe. These interviewees also expressed concern about contamination, particularly regarding well water and "run-off" into residential areas.

***Odor.*** Almost all Tulsa interviewees attributed odor to biosolids land application. Tulsa Landowners were familiar with their neighbors' complaints about odor, but they believed odor was "minor," "short-lived," and "worth it." Tulsa Neighbors cited odor as a significant concern and attributed odor to health, quality of life, and livestock impacts. Odor was the most prominent factor cited by those Tulsa Neighbors who said they were "against" biosolids land application in their county. Odor did not appear to be as significant an issue for VA Neighbors generally, with the exception of some interviewees in one VA county who commented on the impact of the odor on their quality of life and property values.

***Concerns about contamination.*** Interviewees believed this could result in a number of negative health and environmental impacts. In particular, interviewees raised concerns about water contamination from well water or from "run-off" into residential areas. Other concerns included pollution of streams and air pollution.

Some interviewees expressed less concern about biosolids contamination. Most of these interviewees commented that biosolids is a "natural product" and compared biosolids to chemical or animal fertilizers that are applied on the local fields.

***Concerns about the "unknowns" about biosolids.* (e.g., *emerging contaminants and long-term impacts).*** VA Near Neighbors in particular raised concerns about long-term impacts. VDH Officials also expressed concerns about emerging contaminants and the "uncertainty of the science," and focused on the long-term impacts and "non-infectious compounds," such as hormones, medications, heavy metals, and volatile organic compounds.

***Trust in the people who produce and apply biosolids.*** This included interviewees' confidence in the adequacy of the regulations to protect against potential health and environmental impacts. The few Tulsa Near Neighbors who expressed confidence in the safety of biosolids commented that "[the people who are applying it] were doing the right thing." However, most Neighbors in both cohorts commented on their lack of trust in decision-makers, some spontaneously mentioning that they do not trust the government and/or city officials. Some Neighbors mentioned that they do not trust the utility, citing motivation, regulations, and site selection as the key reasons for their lack of trust.

***Agricultural and economic benefits to farmers and to the community.*** Interviewees who said they were "in favor" of biosolids land application in their counties emphasized the economic benefits. Tulsa Landowners cited (and in some cases quantified) numerous direct benefits including richer soil, faster growth of cattle and increased crop production, cost savings on fertilizer, and increased revenue. Neighbors also commented on the benefits, but more in terms of indirect rather than direct benefits. Many believed that biosolids are low cost or free, and that "landowners want it so it must have some benefits as a fertilizer."

Both Landowners and Near Neighbors in all cohorts mentioned environmental and economic benefits to the community, commenting that biosolids land application is a low cost and environmentally friendly way to manage ("get rid" of) a wastewater by-product in a way that benefits the community. One interviewee commenting on the economic benefits to the community noted, "If they did it right, they could reduce everybody's [water/wastewater] bill by selling biosolids to the farmers and thereby subsidizing the wastewater treatment plants themselves." When commenting on the benefits of biosolids, VDH Officials generally referred to them in terms of the "best of both worlds" pointing out the agricultural benefits of biosolids and their role in waste management.

***Other factors affecting judgment*.** These included the perceived source of biosolids (a few interviewees were in favor of biosolids that come from the local communities), perceived use of chemicals in treatment of biosolids (a negative influence), perceived risk of hazardous waste (a negative influence), perceived risk of overfertilization, and impacts to wildlife. Some interviewees compared biosolids to chemical or animal fertilizers that are applied on the local fields. These interviewees generally expressed less concern about biosolids, describing it as a "natural product."

***Findings related to communications.*** Many Near Neighbors, both in Tulsa and in the VA communities, spontaneously commented on the lack of communications by the utilities and companies involved with biosolids. Some specifically mentioned that there was no "advance notice" of biosolids being applied in their neighbor's fields. Interviewees offered several opportunities to improve engagement with people in their communities about biosolids, including being more transparent in decision making, communicating in advance of a land application, providing information on biosolids treatment and application processes, providing balanced information on the benefits and risks, and providing evidence of to support the assurance of safety. Federal and state officials were identified with the most frequency as the people to best communicate about biosolids. Communications from the local authorities or the biosolids organizations elicited mixed responses. Those who favored communications from biosolids professionals commented on their perceived fairness and credibility; conversely, those who expressed more skepticism about biosolids professionals' communications commented on their perceived bias and lack of credibility. A few Neighbors believed farmers should communicate with their neighbors because they have the knowledge and "can speak to the pros and cons."

## Step 4: Develop and Pretest Communications Plan and Materials

With the insight generated from the previous steps, the Biosolids Team was ready to develop and pretest a communications plan and supporting materials. The researchers and the Biosolids Teams looked at what their local stakeholders already knew, what they didn't know or misunderstood that was consequential and, importantly, what these stakeholders said they wanted to know. The Team also considered who the stakeholders trusted and the communications processes they desired. This insight was used develop draft strategies, plans, messages, and materials.

The key "products" that were developed as part of this Step included:

- ***A draft and prototype plan for conducting neighborhood dialogue meetings (Tulsa).*** This plan included the process for communicating with and engaging neighbors in dialogue about upcoming biosolids land application. For example, the plan specified that City staff meet with the landowner 4 weeks before a planned application to confirm the application and discuss the upcoming stakeholder dialogue session (optional: advising the landowner to begin posting "No Trespassing" signs). Three weeks before application, the City staff would send neighbors living within approximately one mile of the planned application site an invitation to the dialogue session, along with the biosolids pamphlet (described later).
- ***Recommendations on the development of guiding principles for the biosolids program (Virginia).*** Guiding principles are a public statement of what the leaders and employees of an organization or association of several organizations stand for and how they want their conduct to be judged by their stakeholders. The

draft and prototypical principles developed for the VA Team focused on earning stakeholders' trust by: demonstrating respect for stakeholders and their perspectives, engaging in open and transparent dialogue, being the first and best source of information about the local land application program, and recognizing that their actions speak louder than words.

- ***Recommendations on defining biosolids partners' role in community engagement (Virginia).*** Because land application in Virginia involves a large set of interdependent partners (e.g., municipal utility managers, contracted land appliers, and the Virginia Biosolids Council), the researchers developed a draft and prototypical template for the VA Team to define the minimum role and potential higher role for each biosolids partner. For example, the Virginia Biosolids Council may maintain its website as a shared resource for all members (at a minimum). Its potential higher role could be to develop communication materials that members could customize for their own communities.
- ***Communications Materials:***
  - Biosolids pamphlet (Tulsa): A pamphlet (Fig. 10.7) on biosolids was developed to provide general information on Tulsa's biosolids program and address the topics identified in the research.

**Q. How are biosolids made?**
Biosolids are produced at both the City of Tulsa's Northside and Southside wastewater treatment plants. First the wastewater is processed and the solids from the wastewater are removed. These solids, called *sewage sludge*, then undergo a number of processes to destroy potential disease-causing organisms such as bacteria, viruses and parasites, called *pathogens*. Once treated, this product is called *biosolids*. The biosolids may undergo further processing to remove excess water. Depending how much water remains, the biosolids may have the consistency of a thick liquid or soil. The biosolids are stored at the wastewater treatment plant until they can be transported and applied on farm land.

*Treatment to remove pathogens*

*Excess water removed from biosolids*

**Q. Why does the City of Tulsa produce biosolids?**
Everyone who lives, works, shops and plays in and around the City of Tulsa contributes to the production of more than 10,000 tons of sewage sludge per year. As part of its wastewater treatment process, the City of Tulsa is required to treat and manage this sewage sludge. Like many other cities and towns across the U.S., the City of Tulsa staff has found that offering biosolids for use on local agricultural land is the most beneficial and economical way to use this recycled product. This helps local farmers and ranchers, avoids the cost and environmental impacts of landfill disposal and avoids the waste of a useful product.

*Storing biosolids at the treatment plant*

**Q. What does the City of Tulsa do with the biosolids it produces?**
Since 1984, our team at the City of Tulsa has provided biosolids to landowners in the surrounding area who have expressed interest in applying biosolids to their land. We select the safest and most economical places to apply biosolids, using criteria such as the distance from residences and water sources, the distance from the wastewater treatment plant, the amount of land available, and the ability of the farmer or rancher to assist in the preparation of the land. Before biosolids can be applied to any land, the City of Tulsa staff must get a permit from the State of Oklahoma. We must show that the land meets strict criteria so we can ensure the safety of people, animals and the environment.

*Crop field of land application vs no application*

**Q. How do biosolids benefit farmers and ranchers?**
Biosolids are used by local farmers and ranchers to improve the soil quality of their land and nourish their crops and pastures. The landowners that we work with have told us that using biosolids results in significant and longer-term improvements in crop yield, pasture quality and rate of growth. Using biosolids lowers production costs for farmers and ranchers by reducing the need to buy more expensive chemical fertilizers. Biosolids make excellent fertilizer because they are rich in nutrients such as nitrogen, potassium and calcium. Biosolids also contain organic materials that replenish the soil and boost its ability to absorb and store moisture needed by the crops to grow. These nutrients and other materials are released into the soil slowly and can be used very efficiently by crops. This means that crops benefit from the biosolids for a longer period of time than they would if treated with chemical fertilizers.

**Q. What are the benefits to the community?**
Applying biosolids on agricultural land benefits the citizens and taxpayers of the City of Tulsa and its surrounding communities. It is a cost-effective and environmentally-responsible way to recycle wastewater by-products and put them to beneficial use. Every year, the City of Tulsa produces approximately 10,000 tons of biosolids – 2,500 truckloads – that are applied on local agricultural land rather than being sent for disposal to Quarry Landfill! This is good for the environment and reduces costs to taxpayers. And, the success of our biosolids program is a demonstration of our local communities' commitment to environmental stewardship.

*Before an application of biosolids*

*6 months after application*



**Fig. 10.7** Excerpt of Tulsa Brochure "The Beneficial Use of Biosolids"

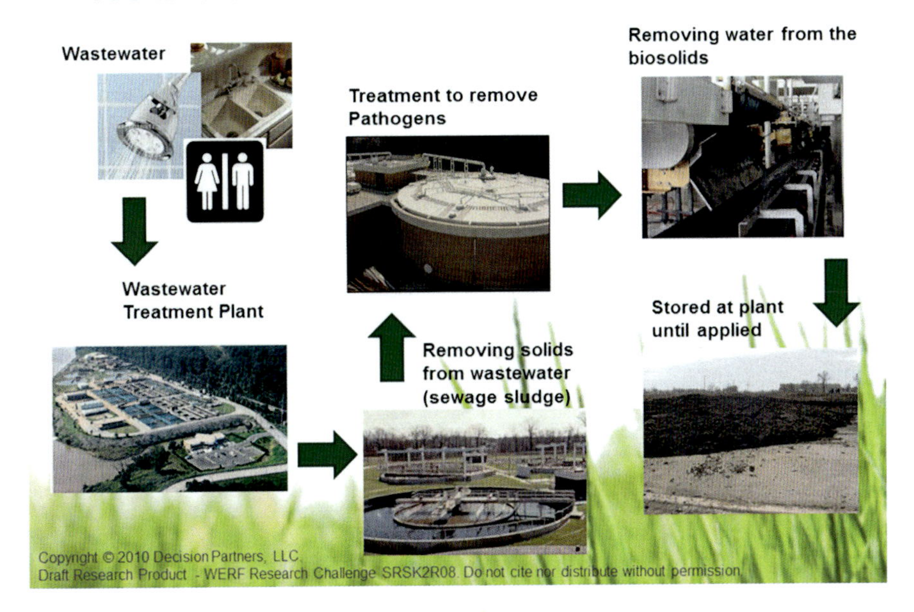


**Fig. 10.8** Excerpt from Tulsa biosolids dialogue presentation

- Generic presentation and speaking notes (Tulsa): A draft and prototypical 20-min presentation (Fig. 10.8) was developed to support communications with neighbors and other key community stakeholders. It provides information related to the key topics that were identified in the research, written in the voice of the Tulsa Biosolids Team and focusing the discussion on the local context.
- Technical FAQ (Tulsa): Through the case study research, more specific questions were raised by the Landowners and Neighbors. The draft and prototypical Technical FAQ developed by the researchers provided Tulsa Team members with additional guidance, references, and draft sample responses to help them address the more specific questions that might arise during dialogue.
- Draft Signage (Tulsa and VA) to post before, during, and after a biosolids land application. The prototypical signage was designed to focus on biosolids as a *product* with beneficial use, and provide useful information about the specific application.
- Website content (VA): The researchers offered evidence-based recommendations to ensure that the Virginia Biosolids Council website provides local community members information that addressed their interests, priorities, and information needs. Specific recommendations included providing the necessary background about wastewater management generally and communicating details about specific projects (e.g., when, where, and where to go for more information).

### *Pretesting Communication Plan and Materials*

Pretesting communication plans and materials is critical to ensure that stakeholders will understand them and that their messages have the intended effect. The researchers worked with the Tulsa Team to conduct pretests in two small workshops (one with Landowners and one with Neighbors, all of whom had participated in the previous research). The purpose of the pretest was to test the Neighbor Engagement Plan (Plan) and the supporting dialogue presentation, dialogue brochure, and site signage. The pretest also provided the Team members with practice and coaching on dialogue skills. The Team revised the presentation and brochure based on participant feedback. The researchers also worked with the Team to develop the Technical FAQ (described earlier) to help them better address detailed stakeholder questions in future dialogue.

## Implementation and Evaluation

With the Plan and supporting material developed, tested, and refined, the Tulsa Biosolids Team was ready to implement the Plan and engage in dialogue through the communication materials. Periodic refinement to the Plan and materials would be necessary, as the Team continued to learn through their engagement with their stakeholders.

The VA Team implemented the project recommendations and tools. Since the project's completion in 2011, the Virginia Biosolids Council redesigned its website to reflect the research learnings and recommendations. Its members also expressed an interest in leveraging opportunities to more effectively communicate with community stakeholders, as recommended in the research.

### *Developing Guidance for Biosolids Professionals*

The case study case study applications validated the customized Strategic Risk Communications Process™ tailored for conducting outreach and dialogue with community stakeholders on biosolids land application. It was clear, however, that biosolids professionals would need practical and actionable guidance on how best to apply the process. Based on the key learnings from the research challenge and the case study applications, the researchers developed a Primer for Biosolids Professionals[7] to serve as a practical handbook for biosolids professionals. It provides the principles of risk communications and guidance on the tasks and tools to use in each step of the Process, along with supporting worksheets and sample materials.

Adoption of the Process as a leading practice requires hands-on, in-depth training, and coaching to biosolids professions on how to operationalize the methods and

[7] Conducting Effective Community Outreach and Dialogue on Biosolids Land Application: Primer for Biosolids Professionals© 2011 WERF and © 2011 Decision Partners, LLC.

tools described in the Primer. Sustained adoption also requires support and resources from their key partners, such as municipal utilities managers, to fully implement the Process and realize its benefits. Continued applications of the Process can generate increasing value and position the Process as a leading management practice for biosolids professionals and municipal planners more broadly.

## Key Learnings and Demonstrated Value

With the development of a systematic Process, actionable tools, step-by-step guidance, and in-depth applications, biosolids professionals should be much better prepared to design and conduct effective communication with their key stakeholders. The Case Study Teams reported significant value in taking proactive steps to improve their outreach and dialogue efforts. They noted that the Process enabled them to gain insight into their stakeholders' priorities and needs, to understand at what level—who, what, and when—to communicate, and to learn how they could improve their engagement.

Feedback from the Tulsa Team demonstrated that they learned that it is a "risky assumption to assume that everything is fine" and that they would "rather be proactive than end up with a crisis down the road." The Tulsa Team saw particular value in the expert model, which they believed provided a comprehensive "systems view" of their operations and could serve as a decision and communications tool, for example, by helping to track a "game plan" to mitigate, resolve, or explain issues, and avoid "crisis mode." Finally, the Team believed that future applications of the Process could enable other opportunities forward (e.g., adoption of new technologies) and reduce the chance of "backsliding" as city officials and supporters change or move on.

The VA Team members also believed that the application and outcomes of the applied Process added value to their current management and communications efforts. As one VA Team member said: "The key take-away for me through this project is that proactive, direct communication is the goal. This research project produced tools, techniques and methods to better communicate the risks and benefits directly to the public... We're going to use that as a framework and a building block to take the tools we've learned from this project and integrate it in to our current communications program."

More broadly, this project demonstrated that biosolids professionals believe that their biosolids land application programs add value to their communities. They want to communicate more openly and effectively about their operations, products, and services with people in their communities, but may lack the skills and tools to do so with confidence.

Over the course of the WERF Research Challenge, it also became clear that a higher value opportunity for biosolids professionals is to continually earn their "license to operate" by demonstrating their strong commitment to their people, their products and services, and the people in their communities. To do so, biosolids professions must actively demonstrate their commitment, their good performance, and their efforts to communicate honestly and effectively. The researchers offered the following recommendations to the biosolids industry, based on the insight generated through this project:

- Redefine how they think and communication about biosolids: as a product, not a waste. In other words, frame biosolids as a valued product that provides a range of benefits to farmers, communities, and society versus biosolids are a waste that society must deal with.
- Ensure that good performance precedes communications. As many industries and companies have learned the hard way, it is almost impossible (not to mention unethical) to communicate one's way out of bad performance. Addressing performance issues—in operations, product quality, or the work of your people—must come before any communications.
- Lead the dialogue on biosolids land application. Biosolids professionals can demonstrate that they are aligned with the values and priorities of their stakeholders by seeking opportunities to open the conversation about the values that are commonly shared (environmental quality, for example), and the higher order benefits that are consistent with these values (community health and well-being and environmental stewardship, for example). They can also demonstrate that they have listened to—and respect—their stakeholders and their stakeholders' interests by openly acknowledging their concerns and perspectives, providing a balanced discussion of benefits and challenges.

The challenges to achieving sustained local community support are not unique to biosolids programs. They are felt by planners and managers who support a wide range of social and environmental projects and technologies, including wind energy development (Wolsink 2000; Devine-Wright 2005), electricity transmission (Butte and Thorne 2005; Vajjhala and Fischbeck 2006), hazardous waste facilities (Kuhn and Ballard 1998), mental health facilities (Takahashi and Dear 1997), correctional facilities (Shichor 1992), and suburban redevelopments (Simonson 1992). The lessons learned in this project serve as a model for any organization, in both public and private sectors, which requires the trust and support of the citizens and decision-makers within the local community in order to assure the viability and long-term sustainability of their operations, products, and services within those communities.

**Acknowledgments** Special thanks to Alan Hais, Retired WERF Project Manager, and Katherine Sousa, Decision Partners, for their contributions to this chapter. Additionally the authors thank the Water Environment Research Foundation (WERF) and for funding support and significant expert contribution to this work.

## References

Beecher, N., Connell, B., Epstein, E., Filtz, J., Goldstein, N., & Lono, M. (2004). Public perception of biosolids recycling: Developing public participation and earning trust. Water Environment Research Foundation (00-PUM-5).

Butte, G., & Thorne, S. (2005). Opening the dialogue on electricity supply mix: Stakeholder consultation report. Ontario Power Authority.

Devine-Wright, P. (2005). Beyond NIMBYism: Towards an integrated framework for understanding public perceptions of wind energy. *Wind Energy, 8*(2), 125–139.

Health Canada. (2006). Strategic risk communications framework for Health Canada and the Public Health Agency of Canada. Minister of Health Canada (ISBN: 0-662-44596-1).

Krogmann, U., Gibson, V., & Chess, C. (2001). Land application of sewage sludge: Perceptions of New Jersey vegetable farmers. *Waste Management Research, 19*, 115.

Kuhn, R., & Ballard, K. (1998). Canadian innovations in siting hazardous waste management facilities. *Environmental Management, 22*(4), 533–545.

Morgan, M. G., Fischhoff, B., Bostrom, A., & Atman, C. J. (2002). *Risk communication—a mental models approach*. New York: Cambridge University Press.

National Biosolids Partnership. (2005). National Manual of Good Practices for Biosolids. Alexandria VA.

North East Biosolids and Residuals Association. (2007). *A national biosolids regulation, quality, end use, and disposal survey final report*. Retrieved January 10, 2011 from http://www.nebiosolids.org/uploads/pdf/NtlBiosolidsReport-20July07.pdf.

Pepper, I. L., Zerzghi, H., Brooks, J. P., & Gerba, C. P. (2008). Sustainability of land application of class B biosolids. *Journal of Environmental Quality, 37*, 1–10.

Robinson, K. G., & Robinson, C. H. (2006). Biosolids recycling: An assessment of public perception and knowledge. *Proceedings of the Water Environment Federation, WEFTEC 2006: Session 11 through Session 20*, pp. 1070–1077(8).

Shichor, D. (1992). Myths and realities in prison siting. *Crime & Delinquency, 38*, 70–87.

Simonson, R. (1992). Managing community opposition to subdivision development. *Real Estate Review, 22*, 82–86.

Slovic, P. (2000). *The perception of risk*. London: Earthscan Publications, Ltd.

Takahashi, L., & Dear, M. (1997). The changing dynamics of community opposition to human service facilities. *Journal of the American Planning Association, 63*, 79–93.

Thorne, S., Butte, G., Ramsay, L., & Chatigny, E. (2006). Strategic risk communications handbook. Health Canada and the Public Health Agency of Canada.

Vajjhala, S. P., & Fischbeck, P. S. (2006). Quantifying siting difficulty: A case study of U.S. transmission line siting. *Energy Policy, 35*(1), 650–671.

Wardell, M. (2010). Proven marketing techniques for biosolids programs, presented at assessing risks and effective communication of land applied biosolids. Retrieved 14 September, 2010.

Wolsink, M. (2000). Wind power and the NIMBY-myth: Institutional capacity and the limited significance of public support. *Renewable Energy, 21*(1), 49–64.

# Chapter 11
# Using Mental Modeling to Systematically Build Community Support for New Coal Technologies for Electricity Generation

**Sarah Thorne and Megan Young**

## Opportunity for New Coal-Based Power Generation Technology

Capital Power Corporation (CPC), formerly EPCOR Utilities Inc. is a power producer headquartered in Edmonton, Alberta. The company develops, acquires, operates, and optimizes power generation from a variety of energy sources, including a multiunit coal-fired electricity generating station and mine.

CPC leaders had long regarded effective stakeholder engagement as an essential business strategy element, one that is key to safe and sustainable operations at existing facilities, as well as the company's business advantage in developing and operating new projects. Perhaps nowhere in the company's fleet can this be better demonstrated than at its Genesee site northwest of Edmonton, Alberta.

In 2008, CPC was presented with a significant business challenge and opportunity—to use new technology to help meet new federal emission regulations aimed to moving Canada toward a path of deep reductions in greenhouse gases. New regulations would see the phasing out of current coal-fired generation technology. Given the abundance of low sulfur coal in the province, there was significant pressure for the Alberta government and industry to look for new technologies to enable coal-based power production to remain an important power option in Alberta.

S. Thorne, M.A. (✉)
Decision Partners, 1084 Queen Street West, #32B, Mississauga, ON, Canada L5H 4K4
e-mail: sthorne@decisionpartners.com

M. Young
Decision Partners, 7 Sceptre Close NW, Calgary, AB, Canada T3L 1X8



M.D. Wood et al., *Mental Modeling Approach*, Risk, Systems and Decisions,
DOI 10.1007/978-1-4939-6616-5_11

**Table 11.1** Sample community questions on air impacts

| Topic: air | Study area: air quality |
|---|---|
| Detailed studies | Community questions from stakeholder consultation: research, workshops, E-CATG meetings, etc. |
| • Assess potential changes in air quality due to construction and operation (emissions, dust)<br>• 3-D dispersion model used to evaluate baseline and predict effects to air quality 50–100 km from plant<br>• Emissions (e.g., $NO_x$ $SO_x$, $CO_2$, particulates, Mercury) compared to specific provincial and federal emission guidelines<br>• EPCOR's current climate change program and effects of climate change on the air quality assessment will be discussed | *Air emissions*<br>• What is going into the air from the current plants at Genesee?<br>– What is released from the specific modules and units of the plant?<br>• Explain what nitrous oxide ($NO_x$), sulfur dioxide ($SO_x$), carbon dioxide ($CO_2$), and particulate is and how it is produced at the plant<br>– Which of these are greenhouse gases?<br>• What air emissions will be released from the proposed new plants?<br>• What are we seeing at ground level in terms of emissions?<br>– Where does the plume typically hit ground level?<br>– What is being deposited and where?<br>– Are there federal regulations around this?<br>*Controlling emissions*<br>• What are the regulatory statutes regarding emission levels?<br>• Can emissions be reduced?<br>*Monitoring emissions*<br>• What does EPCOR do to monitor air emissions?<br>• Where are the monitoring stations located?<br>– How do you determine where to site monitoring stations?<br>• How do you know what emissions are from EPCOR?<br>• Will EPCOR air emissions reports to Alberta Environment be included in the newsletter and made available to download online? |

Coal-based electricity generation presents a unique challenge in Alberta. Low sulfur coal is readily available, easy to access, and a relatively inexpensive source of baseload[1] stable electricity generation required to fuel the province's growing demand for electricity. In 2008 Alberta had an estimated 1000 years' supply of coal.[2]

[1] Power plants that run continuously over extended periods of time are called baseload plants. The power from these plants serves the base demand of the power grid. A power plant may run as a baseload power plant due to various factors (long starting time requirement, fuel requirements, etc.)

[2] http://www.aer.ca/documents/sts/ST98/st98-2004.pdf—Alberta's Reserves 2003 and Supply/Demand Outlook 2004–2013.

In 2008, coal-fired generation made up more than 40 % of Alberta's baseload power generation and a significant portion of CPC's power generation portfolio.

However, air emissions from the coal-fired technology at that time were a large contributor of greenhouse gas emissions in Canada.

At the time, Genesee's three units were emitting approximately 26,000 tonnes of carbon dioxide ($CO_2$) per day, approximately 8 million tonnes per year. The Lake Wabamun area was emitting approximately 50 % of the province's $CO_2$ emissions.

Under federal regulations at the time, coal-fired power plants were required to meet GHG emission standards or retire when they reached 50 years of operation. This meant that 12 of Alberta's 18 coal-fired generating plants would be retired by 2030. Without action, the remaining 6 generating units could continue to emit harmful pollution, reducing air quality and causing negative impacts—in one case until 2061—on human health.

As part of the Alberta Climate Change Action Plan, the government of Alberta announced a $2 Billion (Canadian) fund to encourage industry to invest in new coal technologies and construct the first scale-carbon capture and storage project in Alberta. CPC proposed a near-zero emission Integrated Gasification Combined Cycle (IGCC) electricity facility at Genesee that would demonstrate new technology to deliver improved air quality from the gasification of coal to generate power and the capture of carbon emissions for permanent storage, known as Carbon Capture and Storage (CCS). This project, if successful, would serve as a model for other jurisdictions in North America and beyond.

As users of the first commercial application of this technology in North America, CPC success would require local community support to host the IGCC and CCS demonstration technologies at its Genesee operations. Facilitating this support would require a systematic approach to create a mental model of the new technologies to maximize the positive first reaction and minimize stakeholder questions and concerns.

## Opportunity at Genesee

Stakeholder engagement had been a continuing feature of community communication and relations activities at Genesee since the late 1970s. Initially, activities focused on building and operating the first generating unit at the site, a world-class, mine mouth, coal-fired power plant which came to be known as Genesee 1 (G1). Two other units (G2 and G3) followed. Community engagement had been ongoing since the late1970s and strong relationships formed, particularly with the nearest neighbors.

In 2004, CPC adopted a corporate wide consultation procedure to provide a cohesive process across all projects in the company. The new approach was based on Decision Partners' Mental Modeling approach for stakeholder engagement, which was aligned with and adapted from the stakeholder engagement and risk communications process described in the Canadian Standards Association's (CSA's)

*Q850 Risk Management: Guideline for Decision-Makers.*[3] The CPC Consultation Procedure was an explicit, science-based management system that included the methods and tools needed to "operationalize" the process at each step on every project. It also included a clear commitment to consultation as a business process expressed as guiding principles, known as "Our Commitment to Effective Consultation."[4] The following principles guided all of CPC's stakeholder engagement initiatives:

- Community members must be heard and understood.
  - Projects must align with citizen and community interests.
- No project is unchangeable.
  - It must fit the community and not the reverse.
  - All proposals must be aligned with and take into consideration stakeholder needs, values, and expectations.
- We must be honest, transparent, and accountable.
  - Our project team members will listen, work collaboratively, and be open and honest.
- We must deliver on all commitments that CPC makes.

CPC's systematic approach and long and careful attention to building strong, positive relationships with people in the communities in and around the Genesee area laid the groundwork for the company to build well-informed understanding of and positive public judgment regarding the IGCC and CCS technologies for electricity generation.

## Key Challenges

Achieving community support would require the CCS Project Team to address a number of key challenges. The first was engaging the local community members on a complex project that involved a number of partners, introducing new technologies, and building community understanding and support for these—and doing so within the constraints of fixed government project timelines for funding. The commercial development of IGCC with CCS had never been done before in North America.

[3] This Guideline (subsequently revised in 2009 as Q850-87 (R2009) Risk Management: Guideline for Decision Makers) is also aligned with the US Presidential/Congressional Commission on Risk Assessment and Risk Management Process and the Australian/New Zealand Risk Management Standard. In addition, our work in strategic risk communications is aligned with the International Organization for Standardization's (ISO) 31000 Guidelines on Risk Management (2009), to which Decision Partners provided input.

[4] Capital Power Corporation (EPCOR) Consultation Procedure, 2004.

Processes for gasifying coal for power generation are very different from traditional technologies for coal-fired generation. The Project Team anticipated that people in the community would have significant gaps in their mental models of the potential project components along with a number of questions they would want addressed before they could make a well-informed decision about the acceptability of the project. There were also local equity issues, including the impact of the project on the footprint or local area, primarily the mine permit area. At the time, there were four coal-fired generation facilities in the region, due to the proximity of the coal seam and source of water. The Project Team anticipated that support for a CPC mine permit expansion would be more difficult to achieve than acceptance of the new technologies, as it related to the threshold and equity concerns of the local residents.

The opportunity for the CCS Project Team was to lay the foundation for the advancement of clean coal power technologies through projects at the Genesee Generating Station and Mine. Realizing the full value of these projects would require enabling local residents to make well-informed decisions on a very complex project. Success depended on a science-informed, systematic approach to design, implement, and evaluate effective stakeholder consultation on these technologies and the broader opportunities they presented. The Mental Modeling approach was best suited to the task and readily adopted by the CCS Project Team.

## Project Steps

The proposed Genesee IGCC project comprised new coal gasification, combustion, and electricity generation units, as well as carbon capture and sequestration facilities. An expansion of the adjacent coal mine was also planned to fuel the new unit and provide sufficient fuel for current operations well into the future.

In the development phase of the proposed project, the initial key steps to formulate the stakeholder engagement strategy were as follows:

1. Develop the Opportunity Statement and Guiding Principles.
2. Develop an Expert Model.
3. Conduct Mental Models Research.
4. Hold a Series of Community Advisory Task Group Workshops.
5. Finalize Community Engagement Strategy and Plan.
6. Hold a Community Workshop to Draft the Environmental Impact Assessment (EIA) Terms of Reference (TOR).

## Draft the Opportunity Statement and Guiding Principles

In a facilitated workshop, the CCS Project Team developed the following Opportunity Statement and Guiding Principles for the project. The Opportunity Statement established the scope and objectives of the consultation initiative to be undertaken for the proposed project. The Guiding Principles, which were published in the project materials and on the project website, outlined the standards by which the CCS Project Team would engage stakeholders and on which they wanted their actions to be judged.

## *Opportunity Statement*

The opportunity for the Genesee CCS Project Team is to lay the foundation for the advancement of clean coal power technologies through projects at Genesee. We will realize the full value of these projects and enable informed decision making through focused and coordinated teamwork and effective stakeholder consultation on these technologies and the broader opportunities they present.

## *Guiding Principles*

Though our consultation, we demonstrate our commitment to:

- Proactively seek, engage, and support meaningful consultation on issues and opportunities related to our business, operations, and new ventures. We understand people's interest in understanding our actions and participating in decisions that affect them.
- Demonstrate respect. We treat our stakeholders and all others fairly and with dignity. We respect the cultures, customs, and values of others. We welcome diverse points of view as these are critical to building shared understanding.
- Be responsive to stakeholder interests and priorities. We use consultation to listen to and learn from our stakeholders.
- Ongoing learning and continuous improvement. We measure both the process and the outcome of the consultation process in order to make course corrections as required and to build our internal capacity.

Though effective Consultation processes, we:

- Foster ongoing relationships with key stakeholders through proactive consultation and communication.
- Enable stakeholders to make well-informed decisions about our business, operations, and new ventures.
- Demonstrate to stakeholders that our decisions are prepared with the benefit of their input and are aligned with their interests and priorities.
- Provide open and accessible communication channels that encourage dialogue and enable stakeholder feedback.
- Enable our employees to converse with family and friends about our business, operations, and new ventures in a well-informed manner.
- Minimize the potential for misinformation that can negatively influence the reputation of our company and people and impede our business, operations, and/or the development of new ventures.

As we work on the Genesee CCS Project, we will:

- Be the first and best source of information about our project.
- Enable community members and stakeholders to participate in the decision-making processes by building capacity regarding the various regulatory processes.

## Develop Expert Model

An *expert model* is an influence diagram illustrating an overall system. It summarizes the important knowledge about a topic and illustrates inputs and outputs of the variables within that system. The expert model that was developed for this project, entitled the Expert model of *Influences on Individuals' Judgments on the Acceptability of IGCC with Carbon Capture & Storage* was adapted from a general model of individual's judgments of an electricity system developed in conjunction with EPCOR Utilities Inc. (Fig. 11.1).

The Expert Model served as the analytical framework for the mental models research that followed.

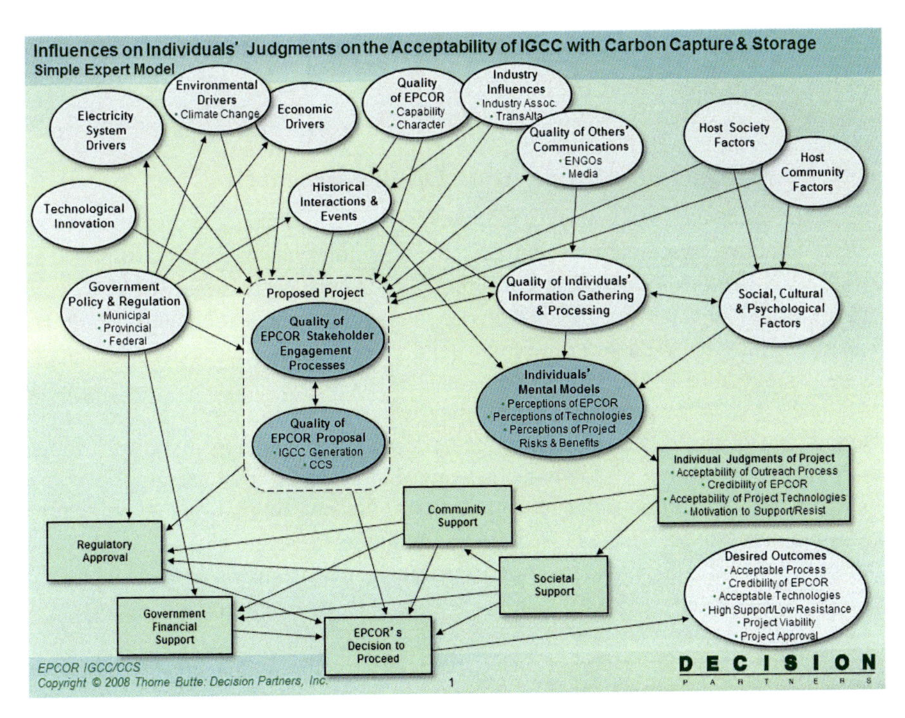


**Fig. 11.1** Simple expert model of influences on individuals' judgments on the acceptability of IGCC with carbon capture & storage

## Conduct Mental Models Research

Formal mental models research was conducted with community residents as the first step to learn about their interests and priorities about the potential project and the technology involved.

For the research, a Mental Modeling approach was used to design a conversational protocol for use with 32 citizens[5] who lived near the Genesee and Keephills Generating Stations. The protocol, which was not circulated in advance, offered interviewees an agenda of topics to discuss in a way that allowed for free expression. They were encouraged to spontaneously raise additional topics and elaborate on their perspectives on these topics. Interviews were tape recorded, with interviewees' permission, to enable subsequent and in-depth analysis, if requested. All of the interviews were confidential. Only aggregate results are reported here. Interviews averaged 34 min in length, ranging from 17 to 56 min.

The interview protocol for the research focused on five topic areas. First, getting initial thoughts on power generation in Canada today and in the future. Next, researchers focused on learning about perceptions related to advanced coal generation technologies, and then more specifically about IGCC and CCS. Final questions during the interview focused on learning about what questions people had about IGCC and CCS and how they would like to be communicated with about the project. The interview ended by asking people what final advice they would give the CPC Team as they moved forward in the project.

## Key Learnings from the Mental Models Research

The community research revealed that 65 % of interviewees would support an IGCC/CCS demonstration project, primarily because they believed the new technology would be better for the local environment and, some added, their health. By extension, they believed it would contribute to improving the environment more generally, specifically air quality and climate change.

The 20 % of interviewees who said they were opposed to the proposed project were primarily concerned about the concentration of power plants in the area, rather than the demonstration of a new technology. The research findings suggested that many of these would be more receptive if they understood that one or two of the older plants in the area were close to the end of their life cycle and would likely be decommissioned. Those reductions, plus the generation from this new, cleaner technology, would result in a significant net reduction in air emissions for the area.

Most of those interviewed (70 %) had concerns, including some who were generally positive about the proposed project, and provided a number of questions that they would want to have addressed. The key issues for those who opposed the

[5] This sample size is typical for a research project of this kind.

IGCC/CCS project were related to threshold and equity for area residents as illustrated by the following quote: "We have four plants in the area now and why here?"

The research also identified that interviewees did not have a complete mental model of either IGCC or CCS technologies. Consequently, they did not have enough information to make a well-informed decision about the acceptability of this project. About half of the interviewees said they had heard about CCS, but their understanding was limited.

- "Isn't that when they separate the gas from the coal and use the gas to generate electricity?"
- "I believe it's something they talk about for Alberta where they light something underground."

Interviewees had several questions that would need to be addressed before they could support a demonstration project in their community.

- "How is it going to affect the community? Are they going to impact the local landowners? I guess the effects on the community would be the biggest [question]."
- "What is the carbon dioxide doing when it's underground? How does the environment deal with it? Does it just sit there or is it going to slowly seep into other places?"

When they were read the description of the potential project, many interviewees (40 %) offered positive thoughts, primarily related to the improved technology and environmental benefits. However, there appeared to be significant variances in perceptions on many topics between the cohorts. These could be indicative of fundamental differences of the different power producers in the region and how they worked with respective communities.

Interviewees also believed that the federal and provincial governments and CPC should share in the cost of the demonstration project, but most noted that ultimately, "the consumer pays."

The research identified the need for the Project Team to help community members learn about these technologies. Additionally, community members had several questions that would need to be addressed before they could support an IGCC/CCS demonstration project in their community (Fig. 11.2).

The results of the research were reported back to the community through a series of newsletters and they were posted on the project website.

## Hold a Series of Community Advisory Task Group Workshops

As the next step in the ongoing outreach and dialogue process, several community workshops were designed to build on the research results. These facilitated mental models-based deliberative dialogue sessions were held in the late summer and fall of 2008 with local residents. The purpose was to provide an overview of CPC's proposed project and elicit input from community members.

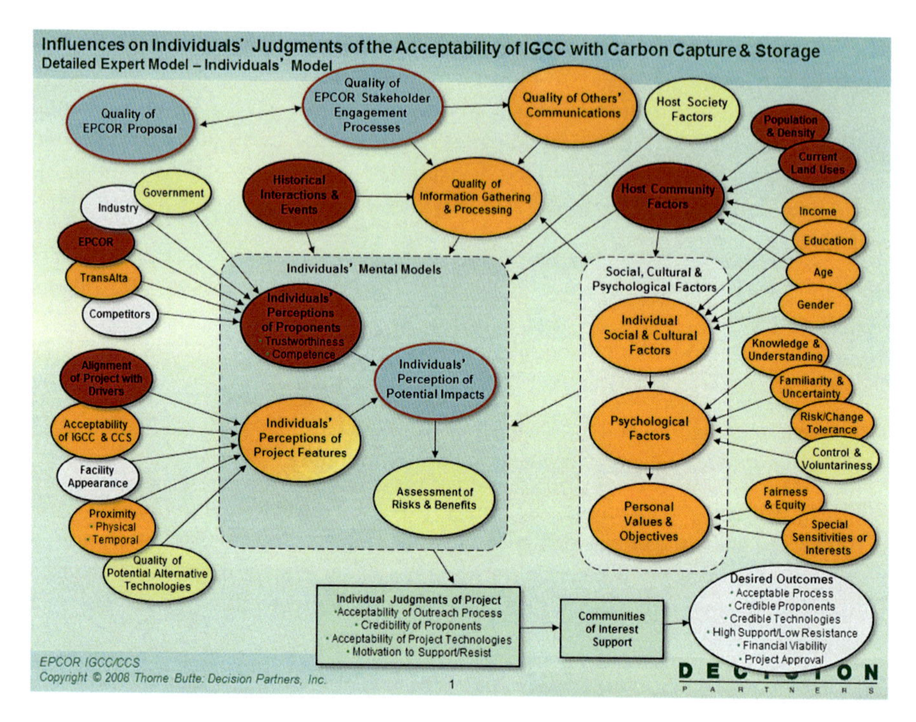


**Fig. 11.2** Detailed expert model of individuals' judgments on the acceptability of IGCC with carbon capture & storage

The purpose of the workshops was to build on the mental models research findings and elicit thoughts and recommendations from community members on how CPC could best design and conduct effective community outreach on the proposed project. CPC invited approximately 30 key community members to attend full-day workshops.

At the facilitated workshops, CPC presented background information on the IGCC and CCS technologies being considered for the project along with an overview of the development and regulatory processes.

Through the process of a deliberative dialogue, participants worked through tough choices to explore the areas of common ground from which alternatives could be developed. Deliberation is a way of discussing important issues and wrestling with tough choices. It enables participants to share their reasoning on issues, talk together, and ultimately work through choices to solutions. In a deliberation everybody has a say and everybody listens. Participants explore what others think, as well as their own beliefs. They don't have to come to conclusions. But they do weigh the consequences of various options based on what is truly valuable to them, and to others.

At the workshops the community members provided the CPC Team with constructive coaching on how to conduct effective communication with the community on the new technologies and on how they wanted to be engaged. There was general support for IGCC CCS, though qualified. Although there were still many questions that needed to be addressed, there was a general openness to these technologies and a desire to learn more through ongoing communications. IGCC was seen to be better than existing technology and community members suggested the new technology should replace existing coal-fired generation plants in the area. For most participants, global climate change was off the radar as—their focus was on improving the local environment, particularly reducing air emissions. Another significant theme was the perceived fairness regarding the distribution of benefits and burdens—the people in the community live with the power plants, and others in the province and beyond, get the power. It was determined that working with community members to determine appropriate "offsets" for the surrounding community members would be critical going forward. Community members also acknowledged CPC's efforts over the past few years to increase and improve outreach efforts. The appreciated the company engaging them in the process at such an early stage.

The results of the community workshops were reported back to the community through a series of newsletters and they were posted on the project website.

## Draft a Community Engagement Strategy

Based on the research results and community input at the workshops, a community engagement strategy for the initial phase of the proposed project was developed by the Project Team as it prepared for the regulatory process. The Team saw the opportunity to further enable informed decision making within the community surrounding the Genesee site about the IGCC project and shape a continuing and constructive community relations program there.

The Genesee community appeared to be open to these new technologies, and generally interested in hosting new technology that would improve the environment, both locally and more broadly. Their support, however, was not unqualified. They had several questions that needed to be addressed about the new benefits and risks associated with the technologies along with the threshold and equity issue of a fifth coal power facility to be built in the region. Addressing these issues would be required to allow the community to have enough information to make a well-informed decision about the acceptability of this project.

CPC's proven Consultation Policy, Procedure, and Process would be leveraged to build understanding of and support for the proposed project. Systematically addressing community interests and priorities, addressing critical questions, and enabling community participation in the formulation of the project opportunity would be critical to success. Managing expectations—both positive and negative—through the preliminary project and technology assessment period would be important.

### Key Elements of the Community Engagement Strategy Included the Following

- A proper introduction of the project opportunity to allow for direct and ongoing dialogue with community members. This could be achieved through one-on-one and small group meetings and workshops with key decision-makers on the project. The focus of this initial dialogue would be to better understand their values, interests, and priorities. Input from these sessions would be used to further refine the engagement plan and develop communication messages and collateral for the project moving forward, based on where stakeholders were at their current thinking or understanding of the project.
- Ensuring consistency on the presentation of the project opportunity. The consultation team for the project was established and each team member received training prior to any engagement on the project. This included a formal consultation team structure with defined roles and responsibilities and regular meetings to ensure stakeholder issues and priorities were included as part of the decision-making process of the greater Project Team.
- Building stakeholder capacity early on to help stakeholders better understand new technologies, regulatory processes, and their opportunities for input to influence project design. As indicated in the research, acceptability of the demonstration technology would require proper framing of the various project components to help build an understanding of the new technologies. Most stakeholders did not have a complete mental model of coal gasification or carbon capture or a clear understanding of how their issues of concern would be reflected in the regulatory review process.

One key opportunity identified to help build this capacity was to work with the local community to draft the Environmental Impact Assessment (EIA) Terms of Reference (TOR). For this project, the development of the EIA was viewed as an important step to work with the community to turn issues into opportunities and to build support and alignment through the process to further reveal mental models—what they know and what they don't know about the proposed technologies.

## Community Workshop to Draft the Environmental Impact Assessment Terms of Reference

As a first step in the engagement strategy, the Project Team needed to help stakeholders understand the new technologies, regulatory processes, and the opportunities for input to influence the project design.

To begin this process, the team identified the opportunity to engage community residents in a workshop for the preparation of the development of the draft Terms of Reference (TOR) for the Environmental Impact Assessment (EIA), a key initial step in the regulatory process for the proposed project.

Typically for an EIA in Alberta, a TOR is developed by the project proponent in conjunction with their environmental consultant to meet the first stage of regulatory requirements for an environmental review. The final draft of the TOR is prepared and made available to stakeholders to provide input before it is finalized with the regulator.

Aligned with the CPC corporate guiding principle, that no project or its elements be unchangeable, the CPC Team decided to take a new approach to drafting the EIA TOR. The approach put the development of the EIA TOR in the hands of community stakeholders. Through a facilitated workshop in collaboration with the Environmental Consultant responsible for the EIA, local community members worked in small groups to develop a document to detailing all of their questions, in their words, to be answered as part of the EIA for the proposed project.

The community input represented a large inventory of questions to be studied or otherwise formally addressed in the course of the EIA and consistent with the CPC team's goal of ensuring the EIA is as complete and comprehensive as possible, with respect to area resident's interests. Questions were grouped, and cross-referenced as appropriate, into the following study areas of possible impacts (positive and negative) with respect to:

- Air Quality—a sample follows
- Water Quantity and Quality
- Land Use and Reclamation
- Noise
- Fish, Wildlife, and Habitat
- Human Health
- Public Safety
- Socioeconomic
- Historical Resources and Traditional Knowledge
- Technology Use and Operation

The EIA TOR that was developed with input from the community for the proposed project required the same environmental studies that CPC would have conducted without community input. The outcome was the same, but with a very different process.

Through the new process of working with the community, in collaboration with the environmental consultant to draft an EIA TOR in their words and questions, CPC was able to build credibility with community stakeholders and the regulators that the project team was committed to addressing key stakeholder issues of concern.

## Finalize Community Engagement Strategy and Plan

As it planned to go forward with the IGCC Genesee Project, the CPC Team used the information and findings from research, workshops, and EIA efforts, as well as continuing consultation initiatives to assure the interests and priorities of people in the surrounding communities and other stakeholders were addressed at every step of the development process and incorporated into considerations on the ultimate project design.

Planned consultation efforts in the next phase (from July 2009 through May 2010) when the project application was filed with the regulator included: presentations to and dialogue with small groups, open houses, meetings of standing advisory groups, meetings of community advisory task groups on specific topics, outreach to municipal leaders and other key groups, formal and informal research, and one-on-one consultation with community residents living within the mine permit expansion area.

Additionally, the CPC Team was readily available to all in the community with an interest in learning more about the project and sharing their views, interests and priorities with Team members.

Specific planned activities included:

- Developing information materials as a resource for communication with area residents. Materials would be written at lay level and built to address key interests and concerns as identified through the research. The focus of the materials would be to help stakeholders understand the new technologies, regulatory process, and opportunities for input to influence project design.
- Coordination of community communication efforts with project partners, to ensure effective and consistent consultation occurs on the part of all project partners. Building understanding with area residents includes creating a consistent approach to communication and consultation from all parties involved in the project.
- One-on-one dialogue, including consultation with individual residents who may be affected by mine permit expansion plans and activities to better understand the benefits and offsets the community would expect in order to make the proposed project more acceptable to the community.
- Hosting public open houses in fall 2009 and spring 2010 to share key learnings from the consultations. Open houses provided the opportunity for the Project Team to demonstrate the company is committed to the community and had been listening to the concerns as expressed to date. The intended outcome of the open houses was to have stronger indicators of what would make the proposed projects more acceptable to community members in terms of possible "offsets."
- Development of a new rural renewal initiative in cooperation with local governments. As identified in the research, community residents raised the concern with a loss of community as a result of the existing power facilities and expanding coal mine. The research presented an opportunity to continue deliberative dialogue with local governments about the opportunities to explore these concerns further and look for opportunities to support renewal initiatives as a community partner.
- Conducting formal community research late fall 2009 to enable a clear understanding of the influences on community judgment of the project before it moves in to the regulatory approvals phase. This second round of research would pro-

vide the community data to indicate whether there was an increased understanding of the IGCC CCS and amine scrubbing opportunities and the company's ongoing commitment to the community.

## Conclusions

Systematic approaches to stakeholder engagement provided CPC with early, evidence-based indicators of community interest and support for the opportunity for Genesee to host the proposed the IGCC CCS facility. While the IGCC with CCS project did not proceed for economic reasons,[6] CPC demonstrated to its community stakeholders its willingness to incorporate their interests and priorities into its projects.

Broadly, effective stakeholder engagement requires:

- *A well-informed understanding, shared by the team, of the total requirements of success*, and why they are the requirements.
- *Up-front, in-depth understanding of stakeholder* values, interests, and priorities—especially their criteria for determining the acceptability of projects.
- *Science-informed, evidence-based process, methods and tools—plus skills.*
- *Full commitment* by proponents to engage in meaningful outreach and dialogue early enough and throughout the process to enable positive business outcomes.
- *Respect for stakeholders.* Listening to and learning from stakeholders is key.
- *Time to build shared understanding* of the project value—perceived need, benefits, and priority balanced against perceived costs and risks.

### Key Learnings

- A research-driven consultation strategy was a key differentiator for the company in the process for Alberta Government funding. The science-based Mental Modeling approach can add value to an organization's stakeholder engagement initiatives by helping to identify people's beliefs, values, and understanding of a proposed project early in the process.
- There is a need to build strategies and communication content based on "where people are at in their thinking today," focusing on explicit decisions and behavioral outcomes. Using a systematic approach to understand people's mental models enables the organization to better identify critical risks early and improve its communication and stakeholder engagement opportunities when siting new projects or new technologies. This approach can significantly reduce

[6] While CPC elected not to proceed with the project, the company was awarded funding for the initial demonstration project as reported in the following news release: http://www.alberta.ca/release.cfm?xID=2638932A61D80-09C7-C9F7-CCD99F5E0113E39A.

the chance of a costly failure in risk communication while improving compliance with best practice.

- The process is as important as the outcome. Evaluating standard regulatory and practices for conducting an EIA identified a gap between how experts and community stakeholders identified issues of concerns. Working with the community in collaboration with the environmental consultant to draft the EIA TOR in their words and questions, the proponent was able to build credibility with community stakeholders and the regulators that the Project Team was committed to addressing key stakeholder issues of concern. The EIA TOR that was developed with input from the community required the same environmental studies that CPC would have conducted without community input. The outcome was the same if not better, but with a very different process and provided a framework for focused engagement as the project EIA progressed.
- To be successful, organizations need to effectively address and sustain stakeholder judgment of the acceptability of projects of long-term, strategic importance.
- Effective stakeholder engagement, based on an in-depth understanding of influences on people's risk-based judgments, is key to minimizing the risk of evoking social friction (local opposition to the project), and managing the social risk throughout the entire project. Stakeholder engagement, including formal consultation, is a strategic undertaking and must be results orientated. Effective and meaningful engagement results in the community being able to make well-informed decisions and actions about a proposed project, enabling the proponent to align the project design to achieve its goals and those of the community. A systematic approach to stakeholder engagement helps ensure the work is done efficiently and progress is unimpeded by unnecessary and costly friction with the community.

**Acknowledgments** This chapter is based on reports from each of the key activities discussed here. Special thanks to Martin Kennedy, Megan Young, Dr. Dan Kovacs, Megan Young, Dr. Laurel Austin, and Anne Papmehl for their contributions to this chapter.

## References

Alberta's Reserves 2003 and Supply/Demand Outlook 2004-2013. Retrieved from http://www.aer.ca/documents/sts/ST98/st98-2004.pdf.

Capital Power Corporation (EPCOR). (2004). Consultation Process and procedures.

# Chapter 12
# Saving Lives from a Silent Killer: Using Mental Modeling to Address Homeowners' Decision Making About Carbon Monoxide Poisoning

**Sarah Thorne, Gordon Butte, and Sarah Hailey**

## Introduction[1, 2, 3]

Carbon monoxide (CO) is a colorless, odorless, toxic gas that is found in combustion fumes such as those produced by vehicles, lawnmowers, stoves, lanterns, burning charcoal and wood, and gas ranges and heating systems. CO from these sources can

---

[1] See: Canadian Standards Association (CSA) Q850-97 (R2009) Risk Management: Guideline for Decision-Makers. This Guideline is also aligned with the U.S. Presidential/Congressional Commission on Risk Assessment and Risk Management Process and the Australian/New Zealand Risk Management Standard. In addition, our work in strategic risk communications is aligned with the International Organization for Standardization's (ISO) 31000 Guidelines on Risk Management (2009), to which Decision Partners provided input.

[2] This chapter is based on the Final Report entitled "Reducing Carbon Monoxide Risk in the Home" (February 2002) and the "TSSA CO Research Executive Summary" (Marc 2002). Special thanks to Sarah Hailey, Research Designer and Analyst at Carnegie Mellon University and former Research Scientist with Decision Partners and to The Technical Safety Standards Association of Ontario, Canada, for permitting us to include this case study.

[3] **Authors' note**: While this case study was completed in 2002, we feel it appropriate to include in this book as it is a robust example of applying Mental Modeling to address fundamental public health and safety issues. CO poisoning and deaths in homes, cottages and boats continue to occur across North America. Carbon monoxide (CO) poisoning leads to an estimated 50,000 ED visits

S. Thorne, M.A. (✉)
Decision Partners, 1084 Queen Street West, #32B, Mississauga, Canada L5H 4K4
e-mail: sthorne@decisionpartners.com

G. Butte
Decision Partners LLC, Suite 200, 313 East Carson Street, Pittsburgh, PA 15217, USA
e-mail: gbutte@decisionpartners.com

S. Hailey, Ph.D.
Institutional Research and Analysis, Carnegie Mellon University,
6608 Dalzell Place #1, Pittsburgh, PA 15217, USA
e-mail: shailey@andrew.cmu.edu



M.D. Wood et al., *Mental Modeling Approach*, Risk, Systems and Decisions,
DOI 10.1007/978-1-4939-6616-5_12

build up in enclosed or semienclosed spaces, and people and animals in these spaces can be poisoned by breathing it. Each year, more than 400 Americans die from unintentional CO poisoning (CDC 2014) and more than 20,000 CO exposures are reported to poison control centers (CDC 2011). In 2005 alone more than 4000 hospitalizations were confirmed cases of CO poisoning and more than 24,000 hospitalizations were suspected cases (Iqbal et al. 2010). Fatality is highest among Americans 65 and older. CO poisoning thus represents a substantial economic and public health burden.

In 2001, the Technical Standards and Safety Authority[4] (TSSA) of Ontario recognized the need develop a research-based risk communication strategy that would improve homeowner awareness of CO safety in the home and encourage homeowners to take appropriate action. Homeowners can take several actions to reduce their risk of CO in the home, such as having their fuel-burning appliances serviced annually by qualified technicians, installing battery-operated CO detectors in the home, and changing the batteries of these detectors once a year. In many jurisdictions, these actions are required by law, including in Ontario, Canada where this study was done. However, in order to take these actions, homeowners must first be aware of the risk, understand its causes and consequences, and understand the appropriate actions that will reduce that risk. The TSSA needed an effective risk communication strategy that would address homeowners' information needs about reducing the risk of CO in the home—one that reinforced what homeowners already knew that is correct, informed what they didn't currently know or misunderstood, and emphasized what they needed to know.

## Communicating the Risk of Carbon Monoxide in the Home

Mental models research provided the appropriate risk communications approach, by first developing a comprehensive model of current expert knowledge and scientific evidence on the risk of CO in the home and how to reduce that risk, and then assessing homeowners' understanding of the risks and the actions they should take to protect themselves. In this case, it was critical to understand what

---

per year in the US (CDC Morbidity and Mortality Weekly Report: http://www.cdc.gov/mmwr/preview/mmwrhtml/mm6030a2.htm). It is also the leading cause of death by poisoning (EM DOCS PRACTICE UPDATES CARBON MONOXIDE POISONING, APRIL 26, 2015. Author: Zach Radwine, MD (EM Resident Physician, University of Illinois College of Medicine—Peoria), Editors: Jennifer Robertson, MD, Lewis Nelson, MD and Alex Koyfman, MD (@EMHighAK)). Understanding what people know, don't know, misunderstand and want know, along with who and what processes they trust, and then using that insight to tailor strategies and communications accordingly, produces measurable outcomes that can potentially save lives.

[4] The Technical Standards and Safety Authority (TSSA) is a not-for-profit, self-funded organization dedicated to enhancing public safety. Headquartered in Toronto, the TSSA enforces Ontario's Technical Standards and Safety Act, 2000 which covers public safety standards in a number of industry sectors including, natural gas, petroleum, propane, and other fuels and equipment, and the certification of technicians who work on fuel burning equipment.

homeowners knew about CO, including the degree to which they associated the importance of annual furnace maintenance with the prevention of a potentially hazardous buildup of carbon monoxide levels in their homes, and what actions homeowners were willing to take, such as having their fuel burning equipment and appliances checked by a certified technician, and repaired if necessary. By assessing the alignments and gaps between expert and homeowner understanding, homeowners' information needs could be identified, and critical gaps in homeowners' understanding prioritized in risk communications efforts. Communications that emphasize what homeowners want and need to know could be designed and tested with new homeowners to ensure that they meet homeowners' information needs. Hence the risk communication strategy is driven by insight into homeowner's values, priorities, and information needs, rather than by guesswork or unilateral decision making.

## Applying the Mental Modeling Research Approach

TSSA, in partnership with the Carbon Monoxide Safety Council, asked Decision Partners to assist in helping to reduce the risk of carbon monoxide poisoning in the home by using the Mental Modeling approach, an optimal approach given its focus on behavioral outcomes and its integral role in recognized risk management processes, including the Canadian Standards Associations' Q850 97: Risk Management Guidelines for Decision-Makers. Mental Modeling was conducted based on a customized version of the six-step Strategic Risk Communications Process™, discussed in detail in Chap. 2:

- *Step 1: Define the Opportunity*—A TSSA Project Team was formed to define the opportunity and scope of the project, review the relevant risk assessment data, and conduct a preliminary assessment of the Team's understanding of stakeholders' perceptions and priorities related to reducing the risk of CO in the home.
- *Step 2: Characterize the Situation*—An Expert Panel consisting of Carbon Monoxide Safety Council members came together in a facilitated expert model workshop to provide in-depth insight required to develop the Expert Model of Reducing the Risk of Carbon Monoxide in the Home—an influence diagram that represented the Council's perspectives on critical factors influencing homeowners' decision making about CO safety in the home. The Model integrated data from previous studies and enabled the expert group to develop hypotheses of the critical gaps (technical and communication) between the experts' and homeowners' understanding of CO risk and personal actions available to minimize the risk. Two primary stakeholder cohorts were identified for the research: seniors living in their original homes and new homeowners.
- *Step 3: Assess Stakeholders' Interests, Priorities, and Communications Needs*—Using the expert model as the guide, we developed a detailed interview protocol for homeowners. We then conducted confidential, one-on-one interviews with 60

homeowners—40 seniors in two different locations and 20 new homeowners. The interviews revealed an in-depth and first-hand understanding of homeowners' perceptions, priorities, and communications needs. The interview data were coded and analyzed against the expert model, and gaps and alignments were identified. This insight served as the foundation for the design of the Risk Communications Strategy. Specific risk management intervention opportunities were also revealed.

- *Step 4: Develop and Pretest Risk Communications Strategy, Plan, and Materials*—Insights from the research results were used to inform the design of a focused Risk Communications Strategy, including the development of key messages and materials. The messages and materials were then pretested with a small number of homeowners to ensure they performed as intended.
- *Step 5: Implement the Risk Communications Plan*—The TSSA Team, with support from the Carbon Monoxide Council, implemented the Communications Strategy and Plan developed in Step 4. They monitored, adjusted and refined the plan and materials as required, based on what they learned through their outreach efforts about stakeholders' interests and priorities.
- *Step 6: Measure Process and Outcomes*—The TSSA Team measured the effectiveness of their outreach and dialogue efforts and determined the degree to which they had achieved the Opportunity defined in Step 1. They also assessed the effectiveness of the Process and made recommendations for future Team efforts as part of their commitment to continuous learning and continuous improvement.[5]

The remainder of this chapter describes in detail how this process was applied.

## Step 1: Define the Opportunity

The first step in the research process was to define the project opportunity by clarifying the scope and objectives of the project. A Project Team[6] was formed that comprised members of TSSA, the Carbon Monoxide Safety Council, and Decision Partners. Council members included risk management experts from TSSA, fuel companies, fuel burning equipment manufacturers and contractors, fire departments, regulators, CO detector manufacturers and regulators, medical experts, and

[5] The TSSA Team members who sponsored this research are no longer with the organization; therefore, we were unable to access files regarding the application of this work in ensuing year.

[6] The TSSA Project Team included the following individuals: Ann-Marie Barker, TSSA; Dr. Wayne Evans, Toronto General Hospital; Scott Harris, Union Energy; Fred Halpern, S-Tech; Mark Hutchinson, Deeth & White Ltd; Cengiz Kahramanoglu, Ministry of Housing & Municipal Affairs; John Marshall, Enbridge; Bob McKeraghan, Canco; Marcia McQueen, Enbridge; Bruce Patterson, ULC; Robert Reid, Riello Canada; Glen Turbrett, CSA; Bill Vale, Lennox; Rob Warren, TSSA; John Wastle, TSSA; Captain Steve Welowski, Toronto Fire; and Don Williams, Enbridge Home Services; plus Gordon Butte and Sarah Thorne, Principals and Partners, Decision Partners.

others all dedicated to reducing CO incidents. The Team established that the Opportunity for this project was:

> To save lives by raising homeowner awareness of the risk of carbon monoxide (CO) in the home and the need to take appropriate action, including conducting annual maintenance of fuel-burning equipment to reduce the risk.

## Step 2: Characterize the Situation

With the Opportunity identified, the next step was to characterize expert knowledge of the risk of CO in the home by gathering and integrating the knowledge and hypotheses of the CO experts into an Expert Model of Reducing Carbon Monoxide Risk in the Home. The Expert Model integrated critical data and expert understanding of the risk of CO in the home, capturing a broad and holistic understanding of the important variables that influence that risk. The model was constructed with the Project Team and additional experts through an iterative process of expert elicitation, generation and revision that expanded expert participants' knowledge, assisted the Project Team's ability to identify important stakeholders, and enhanced the Team's understanding of the necessary steps to implement effective risk communication strategies.

### *Development of the Expert Model*

The expert model is a critical tool for understanding and managing complex issues with multiple stakeholders, such as managing the various factors that contribute to the risk of CO in the home. Depicted as an *influence diagram* of the variables that influence an outcome of interest (in this case, reducing the risk of CO in the home), as well as the relationships between and among those variables, helps people visualize and better understand complex systems, which in turn helps them organize information and coordinate actions more effectively.

Expert models also create an understanding that goes beyond the knowledge of any single individual to form a more complete picture of the issue at hand. The model integrates current scientific knowledge with the diverse knowledge of multiple experts with different backgrounds. The group-based process of developing the model respects the contribution of all experts, identifies key gaps and alignments at the expert level, and can reveal immediate opportunities—for communication, risk mitigation, and other critical research agendas.

Although expert models can be created at any level of complexity and specificity, the process typically begin with two levels: a *base expert model*, a high-level representation of the system that contains primary influences on the outcome of interest, and the *detailed expert model*, a deeper and more thorough representation of the system that contains secondary and even tertiary influences. The *Base and Detailed*

*Expert Models of Reducing Carbon Monoxide Risk* can be found in the Expert Model Narrative section which follows.

A draft of the Base Expert Model for this project was first developed through review of existing scientific literature and with input from the Project Team. Once a draft was complete, a panel of experts was invited to participate in an expert model workshop with the purpose of validating the Base Expert Model and generating the necessary knowledge and insight to develop the Detailed Expert Model. Following the workshop, the Project Team also used the insight generated by workshop participants to develop a set of *Stakeholder Hypotheses* about what stakeholders know that is correct, don't know, misunderstand, want to know, and the sources they trust and the communication channels they prefer. The Stakeholder Hypotheses provided a context for defining the sample and developing the Interview Protocol in Step 3 of the research process.

## *Expert Model Narrative*

The following narrative provides the story line or legend for the nodes and their relationships illustrated in both the Base Expert Model and the Detailed Expert Model (Fig. 12.1).

Except for electrically heated homes, all homes in Ontario have fuel-burning equipment and appliances that are potential *Sources of CO*. Incomplete combustion, caused by an inadequate supply of air and/or an ineffective venting system of fuel-burning equipment and appliances, leads to elevated and possibly unsafe CO concentrations in the home. Fuel-burning equipment and appliances are critical components of the ventilation system of a house. The *Concentration of CO in the Home* as well as the *Sources of CO* may be significantly affected by the entire ventilation system of the house.

The *Concentration of CO in the Home* is also driven by a number of other variables. Because carbon monoxide is an odorless, colorless gas, and is virtually undetectable to the homeowner without sufficient knowledge and instrumentation, *CO Characteristics and Behavior* play a significant role in increasing the risk of dangerous concentration levels in the home. Also, the production of CO depends on the effective performance of the fuel-burning equipment and appliances. *Equipment Performance*, therefore, contributes to the level of *Concentration of CO in the Home* and is influenced by its physical condition as well as *Homeowner Practice* and the *House Characteristics*.

The failure to detect CO in the home will result in *Inhalation of CO by Residents*. How much CO a resident is exposed to is influenced by both the *Concentration of CO in the Home* and the length of the time exposed. Concentration influences the degree to which the resident experiences *Health Impacts*.

A resident can become aware of CO concentrations as a result of flu-like symptoms they develop from inhaling CO. But *CO Detection* can also occur through other means such as through a CO detector alarm which can alert the residents to take appropriate interventions in time to reduce the risk of harm. This type of detec-

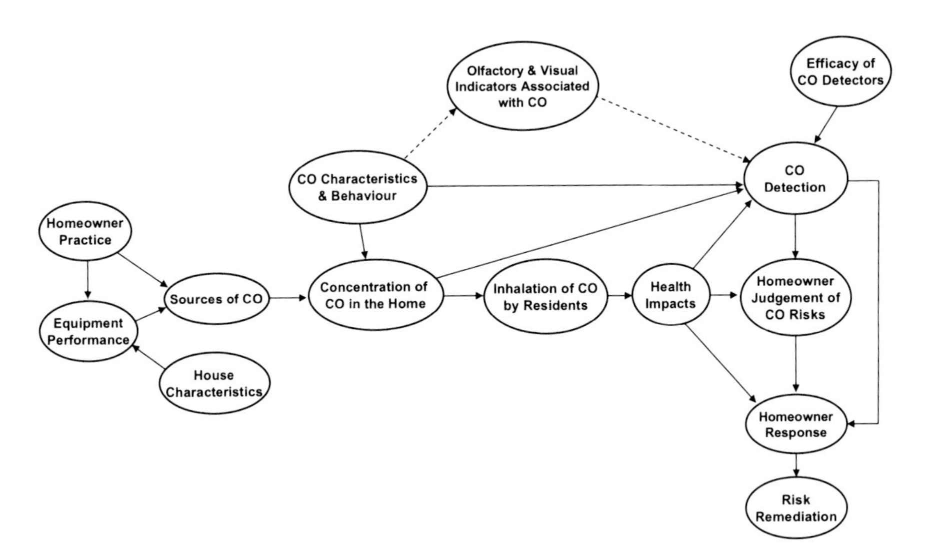


**Fig. 12.1** The base expert model of reducing carbon monoxide risk in the home. Each of the multiple nodes (ovals) in the models represents a variable the influences the outcome of interest. The influence can be positive or negative. An *arrow* between nodes indicates that knowing the value of the node at the *arrow's* tail helps in predicting the value of the variable at the *arrow's head*. It also illustrates that beliefs about the former can influence beliefs about the latter. There may also be a causal relationship between linked variables, or there could just be an "informational" link

tion is dependent on the *Efficacy of CO Detectors. CO Detection*, however, is a reactive measure. It only alarms once there are sufficiently high levels of CO concentration in the home.

Another safeguard against dangerous levels of CO production is through regular maintenance and inspection by industry professionals. They can detect possible CO production before it becomes a high risk and keep equipment operating safely. As they explained, they look for evidence of CO through a number of *Olfactory and Visual Indicators Associated with CO*. These indicators alert professionals to a possible danger, which enables them pinpoint the source and take appropriate actions to eliminate it.

If a high level of CO concentration is detected, responsiveness by homeowners will vary. *Homeowner Judgment of CO Risks* as well as the homeowner's understanding of the *Health Impacts* will shape *Homeowner Response. Homeowner Response* will typically lead to some type of *Risk Remediation* to reduce the risk of CO in the home.

## Expert Model Workshop

Once a draft of the Base Expert Model was complete, members of the Carbon Monoxide Safety Council, plus additional experts, were invited to participate in an all-day Expert Model workshop to review and validate the Model. To structure the workshop, Decision Partners workshop facilitators created a Facilitator's Protocol that organized discussion topics around each node in the Base Expert Model, to ensure that all necessary topics were covered. Because of their broad and diverse expertise in issues related to carbon monoxide, workshop participants had the opportunity to listen to and learn from one other, revealing and clarifying gaps and misperceptions in their own thinking. For example, many experts believed that children were at greater risk of CO poisoning than adults because of their smaller body mass, a misperception that was addressed by the medical doctor on the expert panel. The workshop enabled the Project Team to develop a comprehensive system picture of the risks of CO in the home that went beyond any single expert's understanding.

## Detailed Expert Model of Reducing Carbon Monoxide Risk in the Home

In addition to validating the Base Expert Model, the participants in the Expert Model Workshop discussed reducing the risk of CO in the home in greater depth. This provided the necessary insight for the Project Team to develop the *Detailed Expert Model of Reducing Carbon Monoxide Risk in the Home* (Fig. 12.2), which illustrated a deeper level of influences (Fig. 12.2).

Rather than provide a full Expert Model Narrative of the Detailed Expert Model, describing the issues and influences of each node, we concentrate on the few most relevant to the communications recommendations that were elicited from the research and of which homeowners seemed to be most unaware.

*Homeowner Practice* identifies how the homeowner's behavior can influence the *Sources of CO* and how much CO those sources will produce as a result of incomplete combustion. Homeowner Practice also influences the *Quality of Maintenance and Renovations* in the home, which in turn can affect CO production.

*Homeowner Practice* can be influenced by the following variables:

- *TSSA Regulation for Equipment Maintenance:* the regulation outlining the homeowner's legal obligation to maintain or service his/her fuel-burning equipment and appliances regularly to ensure that they are functioning effectively and safely.
- *Perceived Cost/Benefit of Maintenance:* the homeowner's assessment of the balance of cost versus benefit of regularly maintaining equipment. The *Perceived Cost/Benefit of Maintenance* is influenced by the *Quality of Service Technician.*
- *Quality of Service Technician:* the degree to which service technicians are competent. This includes their ability to assess and deal with a given situation, from maintenance of the equipment and appliances, to identifying hazards, to effectively communicating good practices to their customers. Training, which is mandated and enforced by industry Standards and Regulations, influences the *Quality of Service Technician. Quality of Service Technician* influences the *Quality of Installation* and the *Quality of Maintenance*, as well as the *Perceived Cost/Benefit of Maintenance.*

The *Equipment Performance* node represents the efficacy of fuel-burning equipment and appliances in the home. It is the degree to which these perform properly

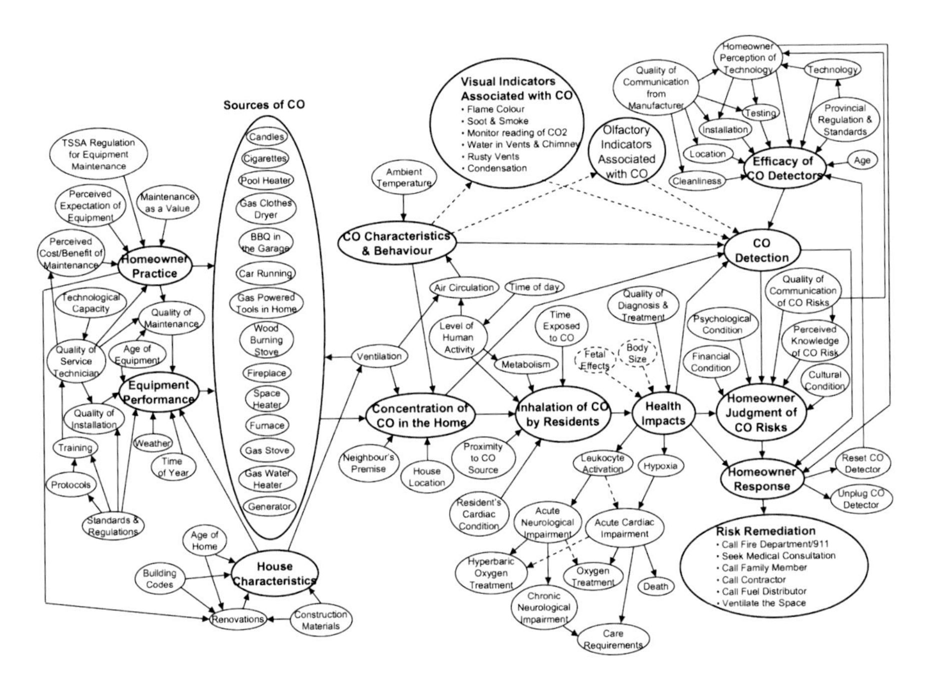


**Fig. 12.2** The detailed expert model of reducing carbon monoxide risk in the home

and efficiently avoiding incomplete combustion. The performance of any fuel-burning equipment or appliance has a direct influence on the *Sources of CO* and the resulting level of *Concentration of CO* in the home.

*Equipment Performance* is influenced by the following variables:

- *Standards & Regulations:* the degree to which industry standards and regulations mandate and enforce both the expectations of the service technician installing and handling equipment, and how the equipment should be performing safely and efficiently. Industry *Standards and Regulations* also indirectly influence *Equipment Performance* by driving safety Protocols, Training, and the Quality of Installation.
- *Protocols:* the degree to which specific safety standards and practices are clearly prescribed for service technicians. *Standards and Regulations* drive protocols. Protocols influence *Training*.
- *Training:* the degree to which industry service technicians are qualified to follow protocols and are encouraged to upgrade their skills so that they are up to date on the latest technology and knowledge. *Standards and Regulations* and *Protocols* drive *Training. Training* influences *Quality of Service Technician.*
- *Quality of Installation:* the degree to which the equipment and appliances are installed accurately according to specifications by a qualified service technician and according to industry *Standards and Regulations*. Thus, *Quality of Installation* depends on *Standards and Regulations* and *Quality of Service Technician.*

### *Expert Model Validation*

Once the Detailed Expert Model was complete, it was reviewed by the experts on the research team and presented to the expert panel at a research forum of Carbon Monoxide Safety Council members. The expert feedback indicated that the Model provided a comprehensive picture of the entire system of relevant influences on CO safety in the home. The Expert Model was then ready to serve as the analytical framework for the design and analysis of the subsequent mental models research with homeowners in Step 3.

## Step 3: Assess Stakeholders' Interests, Priorities and Communications Needs, Through Mental Models Research

Using the Expert Model as a guide, the next step was to design the interview protocol and conduct interviews with target homeowners to gain an in-depth understanding of their interests, priorities and information needs. This understanding could then be compared to the understanding depicted by the Expert Model, illustrating alignments between expert and homeowner understanding that needed to be reinforced, as well as important gaps in understanding how these gaps might be filled.

### *Sample Development*

The sample scope was defined based on TSSA risk assessment data indicating that seniors living in their original homes were at greatest risk, as well as the hypothesis of the Expert Model Workshop participants that new homeowners may not know that they need to have their fuel burning equipment and appliances checked annually by a certified technician. A sample of Ontarian homeowners was developed that consisted of two groups of homeowner cohorts. The first group of "new homeowners" consisted of 20 homeowners from Barrie, Ontario between the ages of 20 and 40 who had owned their home for 10 years or less. Barrie was chosen as it was a rapidly growing community with many first-time homeowners who commuted to Toronto for their jobs. The second group of "senior homeowners" comprised 40 homeowners over 60 years of age, 20 from Sarnia, Ontario and 20 from the Greater Toronto Area. Sarnia was a mid-sized community of approximately 70,000 people, and provided a good contrast to the much larger, more ethnically diverse city of Toronto (now the fourth largest city in North America). Of the 60 total homeowners interviewed, 25 were men and 35 were women. This sample enabled the Team to assess whether new and senior homeowners held different understandings of CO safety in the home and whether there were significant variances in the beliefs and understandings of men and women.

## *Interview Protocol Outline*

A conversational interview protocol was designed for interviewing homeowners in the sample. The protocol consisted of open-ended questions that aligned with the nodes in Expert Model. Key topics focused on eliciting homeowners' beliefs about:

- General knowledge about CO;
- Sources of CO in the home and what causes those sources to produce CO;
- CO detection and the homeowner's response to detection;
- General health impacts from CO exposure; and
- Communication about the risk of CO in the home.

The protocol offered participants the opportunity to speak freely and candidly, as well as raise additional topics and elaborate on those to reveal their individual perspectives in depth. Three pretest interviews were conducted to ensure the protocol questions were interpreted as intended, and insight from the pretests was used to refine the protocol and prepare the analysis.

## *Conducting Interviews*

Potential interviewees were contacted by telephone and invited to participate in a telephone research interview. Because the researchers wanted to discover homeowners' spontaneous mental models of CO in the home, they did not indicate to the potential participants during solicitation that this research was about CO specifically. Instead, researchers indicated that the interview would focus on safety in the home. Potential participants were offered a $30 honorarium for participating and informed that their comments would be confidential to the research team. Interviews were scheduled with participants at their convenience.

All interviews were conducted by telephone. Interviewers followed the interview protocol, asking specific prompts as required. Interviewees were encouraged to respond to questions as they understood them and encouraged to raise additional topics as they came to mind and to elaborate on their perspectives.

## *Coding and Analysis*

The Detailed Expert Model served as the analytical base for coding and analysis. Interviewees' responses were divided into individual segments that represented single ideas, and each segment was assigned one or more codes that best aligned with nodes in the Detailed Expert Model. New codes were created to capture new themes that emerged from the interview but were not included as a node in the original Expert Model. Segments were coded without consideration of the accuracy of the interviewee's statement. Coded segments were then tallied and qualitatively analyzed to assess the relative importance of each factor as related to homeowners' decision making about reducing the risk of CO in the home. Separate analyses were conducted for each of the research cohorts in order to identify potential variations in the perceptions and priorities between senior homeowners and new homeowners, as well as among homeowners in Toronto, Sarnia, and Barrie.

After all segments were coded and assigned to a node in the Expert Model, a *Weighted Mental Model* was created for each cohort that depicted the frequency and intensity of the interviewee's discussion of each factor in the Expert Model. As shown in Fig. 12.3, each node in the Weighted Mental Model was assigned a color that represents the likely strength of influence of that particular factor on interviewees' judgments overall—red nodes represent factors that *many* interviewees raised and are likely to be strong influences on their judgments; orange nodes represent factors that *some* interviewees raised and are likely to be moderate influences, and yellow nodes represent factors raised only by *a few* interviewees. Noncolored nodes represent factors that were raised by experts but not interviewees.

For example, maintenance checks of fuel burning equipment and appliances were a law in Ontario. Further, they did not know that the equipment had to be

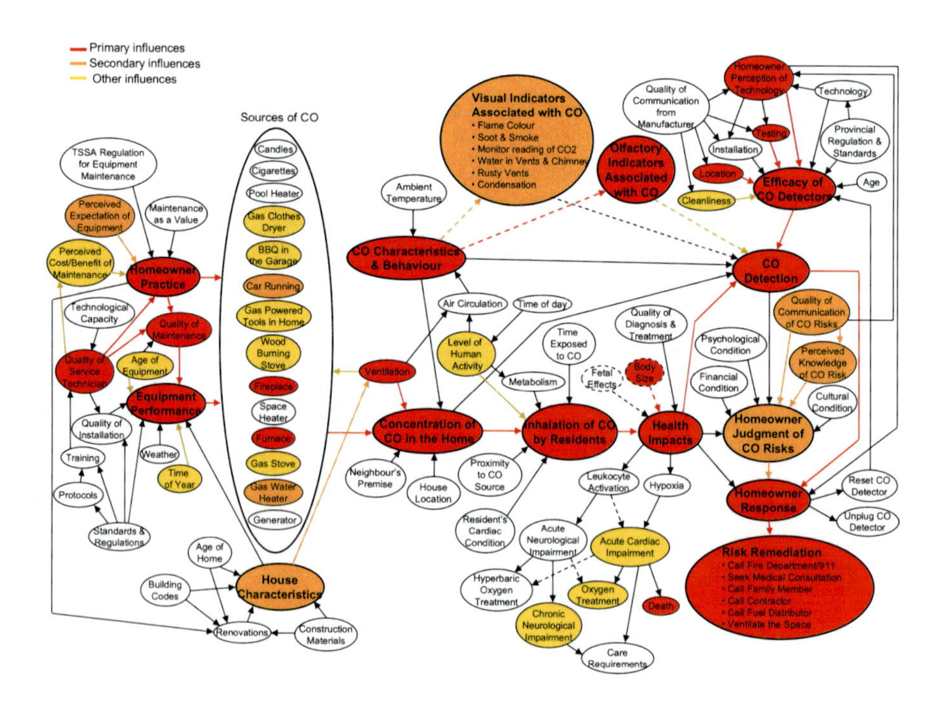


**Fig. 12.3 Weighted Mental Model**

checked by a certified contractor. They did understand that maintenance was important, so building on their mental model to help them understand the frequency required and the training and qualifications of the contractor was an important part of the communications strategy.

Additionally, the research results were compared to the stakeholder hypotheses generated by the Team in Step 2, allowing for an assessment of what homeowners know that is correct, what they do not know or misunderstand that is consequential, what they want to know and who they trust and what communications processes they trust. This allowed the team to identify gaps and alignments between expert and homeowner understanding, and then use this insight to develop prototypical communications plans, messages and materials.

## *Weighted Mental Model*

See Fig. 12.3.

## *Key Results*

### General Homeowner Perceptions

#### Perceptions of Risk of CO

Most people understood that CO was a fatal, "silent killer." They described CO as an "odorless" and "invisible" gas.

A few people, however, had consequential misperceptions about the characteristics and behavior of CO. They thought that CO was "heavy" and lay "close to the floor." A couple of people believed that CO could come from the ground or from sewer gas and could seep into the house.

When considering the overall risk of CO in the home, a few people mentioned that they were uncertain about the overall risk. And a few said that they did not concern themselves with the risks at all. One individual said she was not concerned because her mother worked in the gas industry and answered many "false alarms" with respect to CO. A few people overestimated their knowledge of CO, possibly because of their work history in Sarnia's Chemical Valley.

#### Sources of CO

People identified the furnace and the fireplace as the most common sources of CO in the home. The gas water heater, on the other hand, was one fuel-burning piece of equipment that was often overlooked as a source of CO production. This may be because most often gas water heaters are rented units and homeowners believed that their gas company would automatically service their water heater when necessary.

Approximately two-thirds of the participants understood that damaged or malfunctioning equipment and the lack of servicing and cleaning could lead to the inefficient burning of fuels and increase the likelihood of CO production in the home. Over two-thirds of participants knew that proper ventilation for their fuel-burning equipment and appliances would affect the concentration of CO in the home.

### Maintenance and Servicing

Many people, especially woman, recognized the importance and value of maintaining and servicing their fuel-burning equipment and appliances in order to minimize the risk of CO in the home. More often women mentioned the need to annually service the furnace, while men more often said that the furnace needed maintenance every 2 or 3 years. Men more often had high expectations of their fuel-burning equipment and appliances. They believed that new, "high-efficiency" equipment required less maintenance. A few people commented that they simply did not see the cost/benefit of maintenance and were suspicious of "money grabbing" by contractors.

Most people did not know about regulations requiring homeowners to maintain their fuel-burning equipment and appliances, but over half thought such regulations would be a good idea. Those people who did not think enforcing homeowners to service and maintain their fuel-burning equipment was a good idea were more often men and seniors.

### Service Technicians

Most people reported that they used service technicians to maintain their equipment, as they were the "experts." Some reported that they had "insurance policies" that covered annual maintenance checks by trained technicians.

Approximately two-thirds of participants said that they had never discussed CO with a service technician. Only a few people said that a service technician had spoken directly to them about CO. Indeed, one person believed that a service technician had been misleading about CO and the use of CO detectors. Based on what the service technician told her, she believed that a CO detector would be unable to pick up a reading of CO from her oil furnace.

### CO Detection

Most people said that you could detect CO in the home if you had a CO detector. Three quarters of participants owned at least one CO detector, which was typically located in the bedroom or in the upstairs hallway. Over half of the interviewees said that they were confident to very confident about the CO detector's reliability. However, there appears to be some gap between what people say and do. For example, a couple of people mentioned that they had a CO detector but it was not yet installed.

It appears that most people would respond appropriately if their CO detector went off. They said they would get out, open windows and doors, and call the gas company or fire department. However, a few were confused as to whom they should call and mentioned that they might call a family member. And a few said they themselves would investigate the source and shut it down.

New homeowners more often said that they would call the fire department if they detected CO in the home. Seniors more often said that they would open doors and windows and/or call 911. However, many people were not clear about what action to take first.

Almost two-thirds of participants believed that health impacts, such as "headaches" or "nausea," could be an indication that there might be elevated levels of CO in the home. Most did not indicate what action they would take if they experienced these symptoms.

### Health Impacts

People identified fatigue and nausea as the most common health symptoms associated with CO exposure. Many said that death was the likely consequence of prolonged CO exposure. New homeowners more often mentioned brain damage as an overall health impact of CO exposure. Some participants thought that the overall health effects of CO exposure were respiratory problems.

Participants (and many of the expert panel) did not know that people with heart conditions were at risk from CO exposure in the home. Instead, people believed that children and the elderly were most at risk. They explained that children have smaller lungs and body size, play closer to the floor, and would not know to take appropriate action. The elderly were thought to be at high risk because they were generally weaker than younger individuals. Seniors more often said that the elderly were more at risk from inhaling CO. However, a quarter of the participants believed that CO did not discriminate and that everyone in the house was equally at risk.

### Communication

There appears to be little communication about the risk of CO in the home. One-third of the participants said that they had not received any information on CO risk in the home. About a third said that they had received some information in their gas bill, but they did not remember much of what they read. Despite this, many participants thought that a good means of communication would be "flyers" or inserts with the gas bill. A few participants remarked that they had been prompted to purchase a CO detector as a result of the information they received in their gas bill about CO risks.

High efficiency furnaces and newer technology appear to be sending a message to homeowners that these appliances do not require maintenance, or that they do not require the same level of maintenance as older equipment. This message appears to be reinforced through product advertising and comments made by service technicians. The consequence is that people with high efficiency furnaces, in particular, judge the risk of CO in the home to be minimal.

A few people mentioned that they got information about the risks associated with CO by reading newspaper reports of CO incidents or seeing television reports. They said that the detailed descriptions of CO incidents caught their attention, reminding them of the danger. One person said that he became aware of the risk of CO in the home through television and purchased a CO detector for each of his children and grandchildren. Interestingly, he said that he did not buy one for himself because he sleeps with the window open and therefore did not need one.

If people wanted more information about CO and the risks, many said that they would contact the gas company. Indeed, the gas company was cited as the most common source for information since it was an "informed, credible source."

People also said that they would inquire at the fire department for more information on CO risks in the home. Some people said that they would use the Internet to search for information since it is quick and convenient. However, seniors more often said that they would not use the Internet. They often cited the library as a good source of information as well.

#### Information Requested

Almost half the participants said that as homeowners they would like information about the risk of CO and the health impacts of exposure, especially "that it kills." About a quarter of the participants wanted to know what steps they should take if they detected CO in the home. Some people thought that information concerning CO detectors, such as consumer reports, would be useful. People said that this information could be best communicated through flyers and pamphlets in the gas bill, or through media such as newspapers, radio, and television.

### Comparison of Experts and Homeowner Knowledge

Table 12.1.

### Comparison of Seniors and New Homeowners

- Seniors more often had higher expectations for their equipment performance.
- When asked whether or not there ought to be regulation enforcing homeowners to service and maintain their fuel-burning equipment, seniors more often answered "no."

**Table 12.1** Comparison of expert model and homeowners' mental models of reducing CO risk in the home

| Topic | Experts | Homeowners |
|---|---|---|
| CO description | A deadly gas; lighter than air; odorless, colorless, tasteless | For most interviewees CO was odorless, a silent killer, deadly. A few thought that CO was heavy. A couple thought it could explode |
| Maintenance | Annual maintenance of fuel-burning equipment and appliances. Full service and check | It was important; two-thirds said annual maintenance for the furnace; others said every 2–3 years for the furnace. Check and clean was most frequent description of maintenance |
| Equipment | Damaged, blocked, weather, age of equipment, time of year, house characteristics, and maintenance can all influence the performance | Many recognized that damaged or malfunctioning equipment could affect the performance |
| Concentration | Ventilation affects levels in the home; house characteristics will affect ventilation | Just over two-thirds said ventilation affects levels of CO in the home; some said house characteristics affect levels |
| Health impacts | Flu-like symptoms—headaches, tiredness, nausea, brain damage, death. Only the doctor mentioned cardiac impacts | Over half said fatigue and nausea; over a third said overall health impact was death. They did not mention seeking medical attention |
| Detection | CO detectors; expert assessment; health impacts | Three-quarters said CO detectors; about two-thirds said health impacts. Many have at least one detector |
| Response | Get out, call 911, seek medical attention | Over half said get out; about 45 % said open windows and doors; about a quarter said call the gas company; less than a quarter said call the fire department. None mentioned seek medical attention |

- New homeowners more often said that they would call the fire department if they detected CO in the home. Seniors more often said that they would open doors and windows. Seniors would more often call 911.
- New homeowners more often mentioned brain damage as an overall health impact of CO exposure.
- Seniors more often said that the elderly were more at risk from inhaling CO.
- Seniors more often said that they would not use the Internet as a source of information.

### Comparison of Women and Men

- Women more often raised health issues when describing the characteristics of CO.

- Women more often recognized the value of maintaining and servicing fuel-burning equipment.
- Women more often mentioned the need to annually service the furnace. Men more often said that the furnace needed maintenance every 2–3 years.
- Men more often had high expectations regarding equipment performance.
- When asked whether or not there ought to be regulation enforcing homeowners to service and maintain their fuel-burning equipment, men more often answered "no."

### *Mental Models Analysis*

Once coding and analysis were completed it was possible to perform a mental models analysis—a comprehensive assessment of what the interviewed homeowners knew that was correct, what they didn't know or misunderstood that was consequential, what they wanted to know, who they trusted and what communications processes they preferred (see Table 12.2).

## Step 4: Develop and Pretest Communications Plan and Materials

Insight generated from the research and the mental models analysis enabled the Project Team to develop a focused *Communications Plan* about CO safety in the home that reinforced what homeowners knew and addressed the gaps in their knowledge and their misperceptions. The Project Team first developed the communications goals and the strategic objectives of the communications plan, the key communities of interest as recipients of the communications and potential partners who could assist in maximizing the reach of the communications, the critical message platforms, and actions to be taken by TSSA and the Carbon Monoxide Safety Association (CMSA) to implement these communications. This communication strategy was presented to TSSA's social marketing and advertising agency along with potential measurements of the success of these communications.

### *Communication Goal and Strategic Objectives*

A clear statement of the communications goal and strategic objectives is necessary to define the scope of the Communications Strategy and ensure that the Project Team has a shared understanding of the goals and objectives. The Project Team identified the following Communication Goal:

> To improve homeowners' ability (and that of other Communities of Interest) to minimize risks associated with CO exposure through strategic communication designed to enable well-informed risk decision making on their part.

**Table 12.2** The mental models assessment summarizes what homeowners know that is correct, what they don't know that's consequential, what they might misunderstand, what they want to know, and who they trust

| Interviewees' understanding of CO safety in the home |
|---|
| • *Interviewees Know:* |
| • The furnace and fireplace are sources of CO in the home |
| • Damaged or malfunctioning equipment or lack of servicing will increase the likelihood of CO production |
| • Proper ventilation will reduce the concentration of CO in the home |
| • Service people have the expertise to maintain fuel-burning equipment |
| • The importance of regular maintenance of fuel-burning equipment (a perception held especially by women) |
| • Symptoms of CO exposure are fatigue and nausea |
| • The presence of CO can be detected by a CO detector, or if residents experience such symptoms as headaches and nausea |
| • *Interviewees Don't Know* |
| • Homeowners are required by law to regularly maintain their fuel-burning equipment and appliances |
| • CO affects everyone in a household (as opposed to only those thought to be most susceptible like children and the elderly) |
| • People with heart conditions are especially at risk |
| • CO is a product of all combustion |
| • CO detectors need to be replaced after a period of time |
| • What action to take first if the CO detector alarm is activated |
| • Exposure to high concentrations of CO can cause brain damage |
| • *Interviewees May Misunderstand:* |
| • CO is heavier than air |
| • That furnaces and fireplaces are not the only potential sources of CO in the home |
| • Newer, high-efficiency fuel-burning equipment still requires maintenance |
| • The gas water heater is a possible source of CO in the home. Often, people assumed that since it was a rental, the gas company automatically maintained it |
| • That if the CO detector goes off, homeowners should not seek out the source to shut down the equipment. They should call an expert |
| • What the CO detector is indicating, for example, it is not likely an indicator of a faulty detector |
| • Children and the elderly are most at risk |
| • Respiratory problems are a side-effect of CO exposure, affecting everyone |
| • *Interviewees Want to Know:* |
| • The risk of CO and the health impacts |
| • The steps to take, if it was detected in their home |
| • Consumer reports about various CO detectors and their reliability |
| • *Interviewees Trust:* |
| • *The Gas Company* |
| • *The Fire Department* |

In order to achieve the communications goal, the Project Team identified several strategic objectives:

- Building on the work of the Carbon Monoxide Safety Association, establish TSSA as a catalyst within a network of partners and allied interests for improved CO risk communication.
- Building synergistic communications strategies across the network.
- Developing and pretest communications about CO to ensure accurate and consistent treatment of the subject by all partners.
- Providing model (and pretested) communications tools/ support materials for use by network partners and their employees, as well as with Communities of Interest.
- Maximizing the reach, effectiveness, and impact of the communication initiatives.
- Ensuring the communication strategy is sustainable and maximizes the return on investment (funding and resources) of TSSA and its partners.

## *Key Communities of Interest and Potential Partners*

Communications are most effective when they are tailored to the people who will receive and interpret them. In order to maximize the effectiveness of messaging about reducing the risks of CO in the home, the Project Team needed to identify *Key Communities of Interest*, or communities for whom the messaging would be especially relevant. The Project Team identified several key Communities of Interest on the basis of the research results:

- Seniors who are homeowners.
- New/first-time homeowners.
- Owners of older homes.
- Home appliance/home service suppliers.
- Home appliance store owners.
- Home renovation service providers.
- Cottagers.
- Family physicians; ER physicians; cardiologists; cardiology educators.

Based on the TSSA data, the Project Team identified Seniors who are homeowners as the most at-risk community for the risks of CO in the home. New Homeowners were also a priority cohort based on information provided by contractors and technicians about the lack of knowledge new homeowners have about fuel-burning equipment maintenance. Owners of older homes were identified due to the increased likelihood of possessing older equipment, and cottagers were identified due to their limited occupation of the residence. Individuals with heart conditions were identified as a key group for focused communications, based on the increased risk they face if there is CO in the home. The key Communities of Interest also included people who were mentioned by interviewees and who were therefore likely to interact with these stakeholders, including home appliance and home service suppliers, home appliance store owners, home renovation service providers, family and ER physicians, and cardiologists and cardiology educators.

The Communication Strategy also identified several potential partners with whom TSSA and CMSA could collaborate to maximize the outreach to these Communities of Interest:

- Fire Departments.
- CARP, (Canadian Association of Retired Persons).
- Building Owners Associations.
- Home Builders Associations.
- Home Depot, Home Hardware, Canadian Tire, etc.
- Home Show organization(s).
- Home Renovations magazine(s)/association.
- Cottagers Association/publications.
- CO Detector Manufacturers.
- Appliance Manufacturers Association.
- Gas and oil providers—e.g., Enbridge.
- Service Contractor Association(s).
- Real Estate Association.
- Newcomers' Club; Welcome Wagon.
- Ontario Medical Association (and appropriate subgroups).
- Allied Health Associations, for example, The Heart and Stroke Foundation.
- Ontario Association of Municipalities.

These potential partners were identified during the research as organizations that were uniquely situated to reach out to and engage homeowners. For example, the Project Team identified the Fire Department as a potential partner after learning that during Fire Safety Week (which takes place in October in Ontario), the fire department goes door-to-door to talk with homeowners, presenting an opportunity to discuss CO detectors. The Project Team also identified partners that were specific to the cohorts in the research, including organizations for seniors such as the Canadian Association of Retired Persons, and organizations for new homeowners such as Newcomers' Club and Welcome Wagon.

## *Message Platforms*

On the basis of the mental models analysis and the identified key communities of interest, the Project Team developed specific Message Platforms that were informed by the research results and targeted the specific information needs of the key communities of interest. These message platforms addressed the alignments and gaps

Table 12.3 Message platforms identified by the project team

| Key message platforms |
|---|
| • Remind homeowners that CO in the home is a risk. Homeowners and families should take precautions and be ready to act |
| • All fuel burning appliances in the home produce CO. Maintenance reduces the risk of exposure to harmful levels of CO |
| • Certified service contractors are specially trained to have dialogue with homeowners about CO safety in the home |
| • When the detector sounds, take the right steps. Call 911, leave the house, wait for experts to identify the source and determine when the house is safe to reenter |
| • If you have medical symptoms, see a doctor/get to emergency |
| • Renovate with care. A safe house needs proper ventilation |
| • Cottage safety should include regular maintenance, proper use of fuel burning equipment and appliances, and annual checks for CO |
| • People with heart conditions are at higher risk |
| • Share the mental models (risk communication) research initiative and high level results with CMSA and member organizations and communities of interest |

Informed by the research results, these message platforms reinforced what homeowners knew that was correct and addressed what homeowners didn't know or might misunderstand

between expert and homeowner understanding of reducing the risk of CO in the home as discovered by the research—they reinforced what homeowners know that is correct, while addressing gaps and misunderstandings in homeowners' knowledge. For example, because many homeowners mentioned that they were unaware of what steps they should take if they detected CO in the home, the key message platforms emphasized the need to communicate the necessary steps a homeowner should take when the CO alarm sounds (Table 12.3).

## *TSSA Action Plan*

Once the communications goal and objectives were established, the key communities of interest and potential partners were identified, and the message platforms were clarified, the Project Team was ready to draft an action plan for the TSSA and the CMSA. The TSSA plan recommended broadening the membership of the CO Safety Association to include organizations from the list of potential partners, in order to increase the potential for outreach and dissemination of information related to CO safety in the home. The TSSA plan also provided recommendations for connecting with homeowners at their homes, at the point of purchase of a fuel-burning appliance, and at other locations where they would be likely to visit such as home shows and home repair stores. The plan also included specific strategies for reaching out to the cohorts in the research. For example, reaching out to new homeowners could be accomplished by working through builders, building supply dealers and realtors and new neighbor organizations. The full plan is outlined in the following matrix.

| Channels | Carbon Monoxide Safety Association (with additional partners) role | TSSA role |
|---|---|---|
| CMSA | | • Broaden the scope of the Carbon Monoxide Safety Association (CMSA) to include ongoing communication with key communities of interest, and effectiveness measures<br>• Increase the membership to include professional communicators from existing member organizations plus others on the potential partner list |
| Service contractors certification re: dialogue on CO safety with homeowners | • Provide a small team to assist with the design and testing of the training | • Change requirements of certification to include dialogue with homeowners about CO safety as part of requirements<br>• Commission development of a training module on effective risk communication with homeowners. Ensure the module is pretested with service contractors before it is finalized and made a requirement<br>• Include a simple feedback report so TSSA can track the reach and effectiveness of this intervention<br>• Provide appropriate information materials for service people to leave behind with homeowners<br>• Seek media coverage of the new training requirements for service contractors to enable them to have dialogue with homeowners about CO safety in the home |
| Risk reduction information on CO detectors | • Provide a small team to assist with the design and oversight of this assessment and the development of recommendations for new standards | • Against research findings, assess information on CO risk reduction in the home that is currently provided with the purchase of a CO detector<br>• Commission the development and testing of appropriate risk reduction information—e.g., Stickers, labels, package inserts, package labels/instructions for product use<br>• Work with appropriate authorities to change the standards regarding risk reduction information with CO detectors |
| Information regarding fuel-burning appliances | • Provide a small team of appliance manufacturer members and others to assist with the design of this meeting and pretest the concept with key people in their companies | • Hold a meeting with appliance manufacturers and their marketing/product managers to advise them of the research findings and possible implications, including legal, regarding the "no maintenance" perceptions people appear to have resulting from their marketing efforts to date<br>• Seek opportunities with the group to address this misperception and reinforce CO safety in the home |

(continued)

| Channels | Carbon Monoxide Safety Association (with additional partners) role | TSSA role |
|---|---|---|
| Consistent materials | • Participate in the process to develop and pretest core messages and materials<br>• Define communications plans to maximize the use and distribution of these messages and materials through new and existing channels—for example, mailers in bills to cottagers and homeowners in the north, mailers in bills from gas and oil companies, articles in association publications such as CARP, posters at fire stations, community centers, etc.<br>• Define plans to maximize the distribution of the home safety kit | • Kick off the new CMSA (and its expanded membership) with a workshop to:<br>• Background the members on the principles of effective risk communication<br>• Share research results<br>• Share draft strategies resulting from the research<br>• Hold breakout sessions with similar groups (e.g., appliance manufacturers) to get input into how they could support the strategies, building on their existing communication and outreach efforts<br>• Share work undertaken by TSSA on development and pretesting of core messages and materials, including a pamphlet, bill stuffer and one-page ad<br>• Seek participation by CMSA members in the development of core materials<br>• Seek participation by CMSA members in the development of a cost-effective CO home safety kit<br>• Determine next steps, including a meeting schedule for the group |
| Outreach to homeowners | • Develop a plan to maximize outreach opportunities to homeowners in their homes, at venues where they would be interested in talking about CO safety in the home, and at point of purchase of fuel-burning equipment, building on efforts already in place by members. For example, add questions about CO safety to door-to-door checks conducted by Firefighters during Fire safety week | • Sponsor CMSA (with extended membership) quarterly meetings to develop plans and assess progress |

(continued)

| Channels | Carbon Monoxide Safety Association (with additional partners) role | TSSA role |
|---|---|---|
| Advisory processes | • Provide small teams to assist with the design and implementation of these advisory processes<br>• Publicize these advisory processes and their results in member organizations' publications and on their Websites | • Host an advisory process with people from different social and safety agencies (possibly cosponsored by the Fire Department) to determine how to best communicate about CO safety in the home with people of different ethnic backgrounds<br>• Host a northern and rural advisory process to determine how best to communicate about CO safety in the home and at the cottage with people representing Communities of Interest (including Aboriginals) in rural communities, northern communities and cottagers. This could be cosponsored by the Cottagers Association and could be covered in publications such as Cottage Life<br>• Host a similar advisory process with seniors and agencies who support seniors living in their homes to determine how best to communicate about CO safety in the home for seniors. This could be cosponsored by CARP and covered in their publications |
| Print, radio, and television publicity | • Provide input into appropriate publications<br>• Provide background on relationships with editors and journalists that TSSA could build on<br>• Seek opportunities to incorporate CO safety messages into CMSA member publicity plans<br>• Provide experts from different backgrounds with materials to participate in radio or television panels during home shows, etc., and for interviews with the media on CO safety in the home<br>• Use member publicity channels and publications to let others know that contractors must be certified and are now trained to communicate with homeowners about CO safety in the home—"Ask your contractor about CO safety in the home" | • Develop a plan for seeking publicity—paid space advertising and editorial—in magazines such as Home Builders, Renovations, Cottagers Magazine, CARP's publication, etc.<br>• Work with CMSA to determine the best way to gain publicity to promote CO safety in the home. Test concepts such as a CO Safety act—"Mr. Home Safety"—to do demonstrations and generate interest at home shows and other public events<br>• Seek publicity on the new contractor certification requirements and the training contractors will now require. Promote having homeowners ask their contractors about CO safety in the home<br>• Continue to publicize reminders to check appliances in the fall<br>• Continue to use PSAs to publicize CO safety in the home messages<br>• Continue to publicize incidents and CO safety reminders |

(continued)

| Channels | Carbon Monoxide Safety Association (with additional partners) role | TSSA role |
|---|---|---|
| Outreach to physicians | | • Work with Dr. Evans to determine how to best get appropriate CO safety information to physicians, particularly family physicians, ER physicians, cardiologists, and cardiology educators, for their use and for use by their patients |
| TSSA Website | • Members customize CO safety materials for their Websites. Link to TSSA and CMSA member sites | • Provide CO safety information on the TSSA Website. Link to CMSA members and others alliances |
| Correspondence | • Send research executive summary and high level plans to key communities of Interest | • Research executive summary and high level plans to key communities of Interest |
| Technical and Association Publications | • Members in the appliance and CO detector industries periodically assess technical research on best available technology<br>• Fire Department share project process, research and results at conferences, through trade publications, etc. | • TSSA publish project process, research and results—technical and communication, in:<br>• Industry journals<br>• Safety journals<br>• Technical society publications |

## *Implementation and Evaluation*

With the mental models analysis completed and the communications plan in place, the Project Team presented the recommendations to the TSSA Communications Agency, who used the research results to develop an advertisement campaign, communications materials to be distributed to potential partners, and other risk communications materials. The Project Team also provided several potential measurements of success to evaluate their communications efforts:

- Reach and breadth of safety communication messages through TSSA, partners and alliances, plus dollar value.
- Intensity and frequency of communications.
- Amount of unpaid publicity through magazines, etc., plus dollar value.
- Follow up mental models research, using 2002 research as a benchmark.
- Dialogue and feedback from key Communities of Interest.
- Annual research to track how many people have their furnaces and other appliances serviced annually.

- Feedback reports on dialogue with homeowners from certified service people.
- Ultimately, zero CO-related fatalities and CO-related incidents.

## Key Learnings and Demonstrated Value

Carbon Monoxide in the home poses a risk to people and animals. The Mental Modeling approach provided the TSSA with the research-based risk communication strategy that could be used to improve homeowner awareness of CO safety in the home and save lives from this silent killer. The resulting communication strategy took advantage of a comprehensive understanding of existing expert knowledge and scientific evidence, as well as insight into homeowners' current knowledge and information needs.

The research revealed several research priorities and communication opportunities. For example, the finding that two-thirds of homeowners said they had never discussed CO with a service technician revealed the opportunity to train service technicians to have dialogue with homeowners about CO risks. Because homeowners were often uncertain what steps they should take if they detected CO in the home, they would benefit from information outlining the basic response steps in a format that can be readily available in the home.

The research also identified what information homeowners wanted to receive about reducing the risk of CO in the home. Almost half the interviewees said that as homeowners they would like information about the risk of CO and the health impacts of exposure, especially "that it kills." About a quarter of interviewees wanted to know what steps they should take if they detected CO in the home. Over 20 % of interviewees thought that information concerning CO detectors, such as consumer reports, would be useful. Interviewees thought this information could be best communicated through flyers and pamphlets in the gas bill, or through media such as newspapers, radio, and television.

## References

CDC. (2011). Carbon monoxide exposures—United States, 2000-2009. Retrieved from http://www.cdc.gov/mmwr/preview/mmwrhtml/mm6030a2.htm.

CDC. (2014). QuickStats: Average annual number of deaths and death rates from unintentional, non–fire-related carbon monoxide poisoning, by sex and age group—United States, 1999–2010. Retrieved from http://www.cdc.gov/mmwr/preview/mmwrhtml/mm6303a6.htm.

Iqbal, S., Clower, J. H., Boehmer, T. K., Yip, F. Y., & Garbe, P. (2010). Carbon monoxide-related hospitalizations in the US: Evaluation of a web-based query system for public health surveillance. *Public Health Reports, 125*(3), 423.

Canadian Standards Association. CAN/CSA—Q850-97(R2009) Risk management: Guideline for decision-makers.

# Chapter 13
# U.S. Census Bureau Integrated Communications Services for Data Dissemination: Mental Modeling Case Study with Key Internal Expert Stakeholders

**Daniel Kovacs and Sarah Thorne**

## The Opportunity

The Census Bureau is a complex organization responsible for conducting and making available the results of the Decennial Census as well as hundreds of individual surveys in an increasingly tight budgetary environment. The Communications Directorate within the CB is a key agency responsible for supporting these activities through communications and engagement with internal and external stakeholders, including: Congress (which funds CB activities), program areas within the CB that conduct surveys, other government agencies which sponsor individual surveys or conduct related research, and businesses and individuals who can use Census data and are also potential respondents to Census surveys and the Decennial Census.

In response to these challenges, the Communications Directorate asked our Project Team, led by Reingold in partnership with Decision Partners and Penn Schoen Berland, for support in establishing a customer-centric and data-driven method of communications to better support the CB's mission to make data and analyses from its surveys and censuses available to the general public and other key audiences. In the first stage of this project, the Team developed a Communications Research and Analytics Roadmap to provide the needed insight to guide the development of effective integrated communications services. The foundational mental models research reported here provided the critical insight required to develop the Plan.

D. Kovacs, Ph.D. (✉)
Decision Partners, 1458 Jersey Street, Lake Milton, OH 44429, USA
e-mail: dkovacs@decisionparters.com

S. Thorne, M.A.
Decision Partners, 1084 Queen Street West, #32B, Mississauga, ON, Canada L5H 4K4
e-mail: sthorne@decisionpartners.com



M.D. Wood et al., *Mental Modeling Approach*, Risk, Systems and Decisions,
DOI 10.1007/978-1-4939-6616-5_13

The first step in the Mental Modeling approach was to develop a clear opportunity statement. The following Research Opportunity was developed in collaboration with the CB Team at the kickoff meeting for Task 2[1] in October 2012:

*The research opportunity in Task 2 is to leverage existing research and CB knowledge to develop a science-based, research-informed Communications Research and Analytics Roadmap, starting with the development of an expert model, to guide development and implementation of integrated communications services for data dissemination. We will:*

- *Identify key internal stakeholders' mental models;*
- *Define external audience segmentation;*
- *Identify key internal and external influencers;*
- *Identify key communication channels;*
- *Refine messaging; and*
- *Optimize communications products and marketing materials.*

*Implementation of the Plan will enable the CB to design and implement an integrated communications program that is science-based, customer-centric, and data-driven.*

## Mental Modeling Approach

To gain the insight required to develop the Plan, we applied Mental Modeling Technology™ (MMT™). MMT is an evidence-based and scientifically informed process for understanding and influencing judgment, decision making, and behavior. The science-based Mental Modeling research approach is a critical component of MMT. It was particularly well suited to addressing the challenges described by the CB Team and was well aligned with the customer-centric and data-driven objectives of the project (Chapter 2).

The multiphase Mental Modeling approach comprised the following steps:

- A draft *Base Expert Model of Influences on Integrated Communication for Data Dissemination* was developed based on review of background materials and preliminary discussions with a select group of expert CB employees. This Model was used to design the interview protocol for the expert interviews conducted in this stage of the research. The Base Expert Model was revised based on the results of these interviews and shared with the CB Project Team at a validation workshop on February 12, 2013.
- Drawing on the expertise of key internal CB Project Team members and expert stakeholders, a Detailed Expert Model of "Influences on Integrated Communication for Data Dissemination" was developed. This Model characterizes desired outcomes for effective CB data dissemination and illustrates key factors that drive and influence success of these objectives. This Model enabled us to identify core issues and opportunities for research to be conducted in the second phase of work.

[1] Task 2 was one of several tasks in a larger project that was awarded to Reingold in 2012. Decision Partners' contractual involvement in this project was limited to this Task.

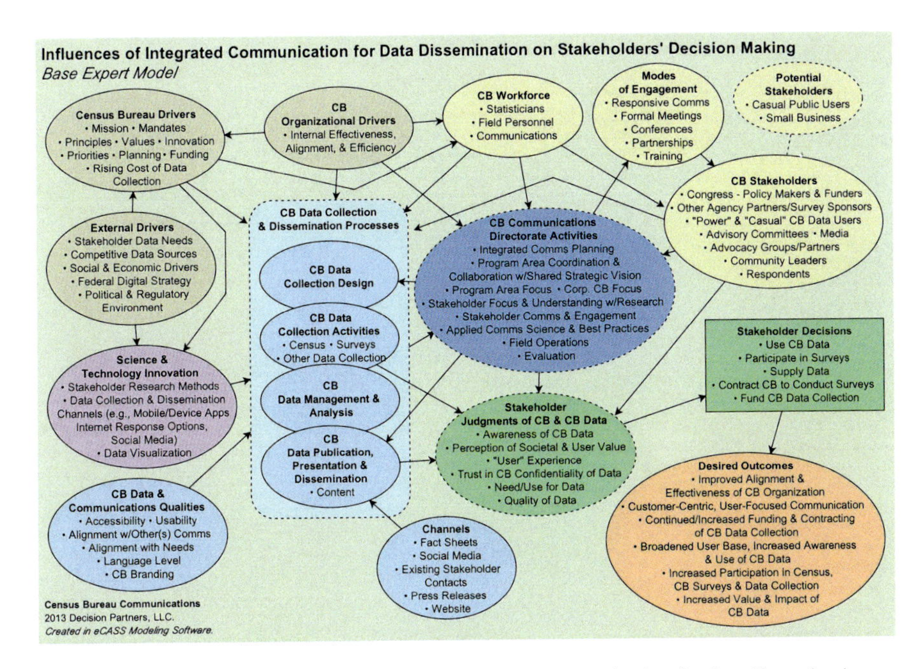


**Fig. 13.1** Base expert model of influences of integrated communication for data dissemination on stakeholders' decision making

- Mental models interviews were then conducted with 26[2] internal CB expert stakeholders to gain further insight into CB activities, stakeholders, and stakeholder engagement and communication activities.
- Based on critical insights from the results of the mental models interviews, the foundational CRAR was developed. Its primary purpose was to support the Communications Directorate as it worked to integrate communications services with the goal of improving public support of, and participation in, CB data collection and dissemination activities.

## *Developing the Expert Models*

A critical step in the Mental Modeling is to develop an expert model. An expert model is an essential management tool, typically depicted as an influence diagram, which summarizes the relevant knowledge about the complex issues being evaluated.

We began by developing a *Base Expert Model of Influences on Integrated Communication for Data Dissemination* (Fig. 13.1) based on our review of background materials and preliminary discussions with a select group of CB employees. This Model was used to design an interview protocol for the expert interviews conducted

[2] This sample size is typical for a project of this nature.

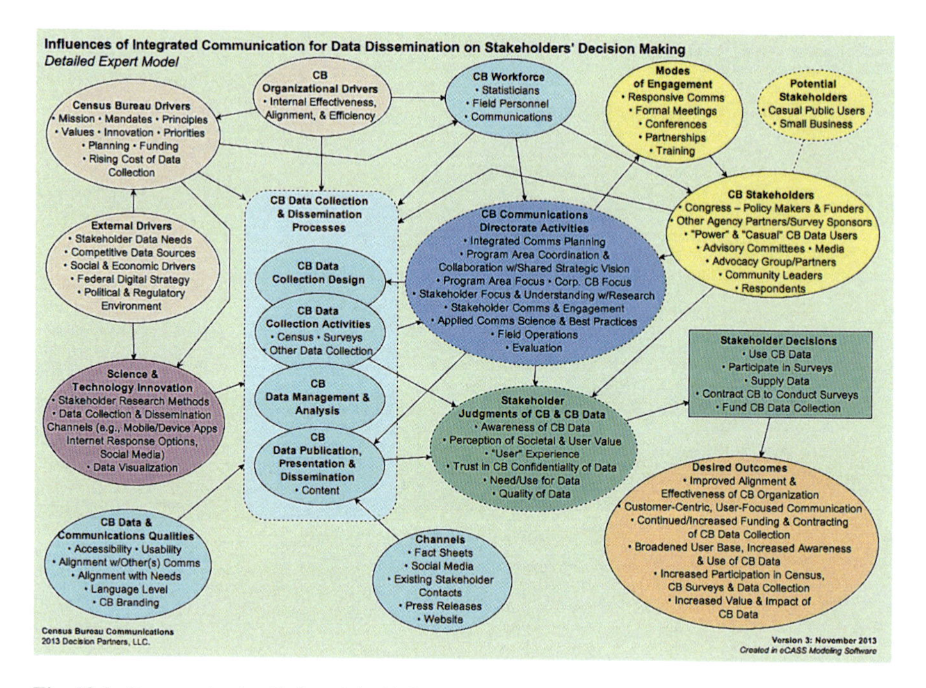


**Fig. 13.2** Presents the detailed model of influences on integrated communication for data dissemination developed based on the results of research presented later

during this stage of the research. The Base Expert Model was revised based on the results of these interviews and was shared with the Census Bureau Project Team at a validation workshop on February 12, 2013.

Drawing on the expertise of key internal CB Project Team members and expert stakeholders, a *Detailed Expert Model of "Influences on Integrated Communication for Data Dissemination"* was developed. This Model characterized desired outcomes for effective CB data dissemination and illustrated key factors that drive and influence success of these objectives. This Model enabled us to identify core issues and opportunities for research to be conducted in the second phase of work (Fig. 13.2).

## *Developing the Sample*

The sample of interviewees, developed by Communications Directorate staff, was designed to provide insight from a broad cross-section of the CB. It included experts from: the Communications Directorate Center for New Media and Promotions and the Office of Congressional and Intergovernmental Affairs; Economic Directorate; American Community Survey (ACS) Office; the Decennial Management Division; the 2020 Research and Planning Office and experts from the Application Services Division; Policy Coordination Office; Research and Methodology Directorate; Geography Division; Center for Survey Measurement; and the Social, Economic, and Housing Statistics Division. Interviewees from the Headquarters Field Office and the Regional Offices were also included to provide on-the-ground perspectives.

### *Conducting Mental Models Research*

The interview protocol was designed based on the Expert Model. Twenty-six in-depth, one-on-one mental models research interviews were conducted with a number of internal CB expert stakeholders to gain further insight into CB activities, stakeholders, and stakeholder engagement and communication activities. The sample of interviewees, developed by the Communications Directorate, was designed to provide insight from a broad cross section of the CB.

The one-on-one, confidential interviews were conducted over the phone between December 12, 2012 and March 14, 2013. Interviews ranged from 19 to 79 min and averaged 48 min in length. Interviewees were asked to respond to a series of mostly open-ended questions on the following topics:

- General background on their responsibilities and expertise;
- CB stakeholders with respect to data collection and data dissemination;
- Modes of communications and engagement with stakeholders;
- What is working well, and what is not working well with respect to stakeholder communication and engagement;
- Opportunities for improving stakeholder engagement and communications;
- The role of the Communications Directorate in supporting stakeholder communications and engagement and strategic alignment with Program Areas; and
- Research needs to inform and improve communications strategies.

## Research Highlights

The following are highlights of results from the mental models research with internal CB expert stakeholders, focusing on the earlier core topics most related to development of the CRAR. The in-depth information from the expert interviews provided critical insights into the development of the CRAR. Interviewees were extremely open and candid. They very much appreciated the opportunity to provide their perspectives on challenges, issues, and opportunities for improvement that they believed were important, and several complimented the research approach.

### *Key Census Bureau Stakeholders*

Interviewees identified a number of key stakeholders, with most focusing on: Congress as a funder of CB activities and as a conduit for questions and complaints from the public regarding CB activities; other government agencies such as the

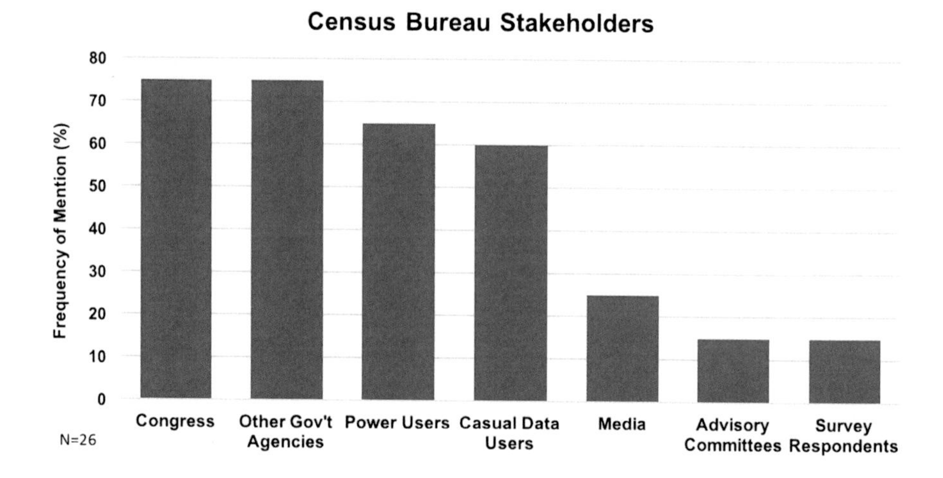


**Fig. 13.3** Census Bureau stakeholders mentioned by interviewees

Bureau of Labor Statistics and Bureau of Economic Analysis who sponsor some CB data collection activities; power users or sophisticated users who use CB data in a commercial or professional capacity such as large business, academics, and advocacy groups; and casual users, individuals could use CB data for noncommercial purposes but are less likely to be aware of data products and less familiar with how to access CB data, including school teachers, small business owners, and the general public.

After a general discussion of their responsibilities and expertise, interviewees were asked to identify and discuss key CB stakeholders associated with the data products with which they were most familiar. The following bar chart summarizes the most frequently mentioned external stakeholders (Fig. 13.3):

## *Modes of Communications and Engagement*

When discussing how the CB communicates with and engages stakeholders, interviewees mentioned a number of modes. The top modes are highlighted in decreasing frequency of mention in the following bar chart (Fig. 13.4):

## *What Is Working Well*

In terms of what was working well with regard to CB stakeholder communication, Interviewees most frequently mentioned: relationships with power users who have structured and ongoing working relationships with CB personnel and higher motivation to find and understand CB data products, planned and proactive communications with

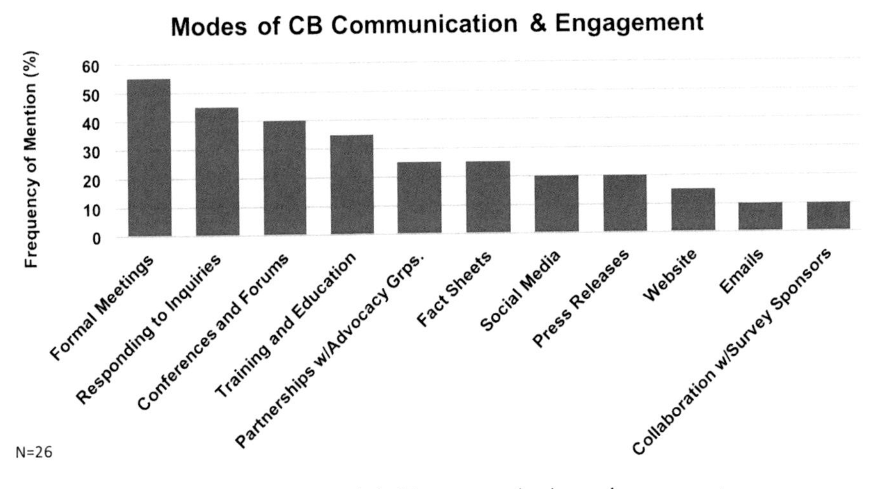


**Fig. 13.4** Modes of Census Bureau stakeholder communication and engagement

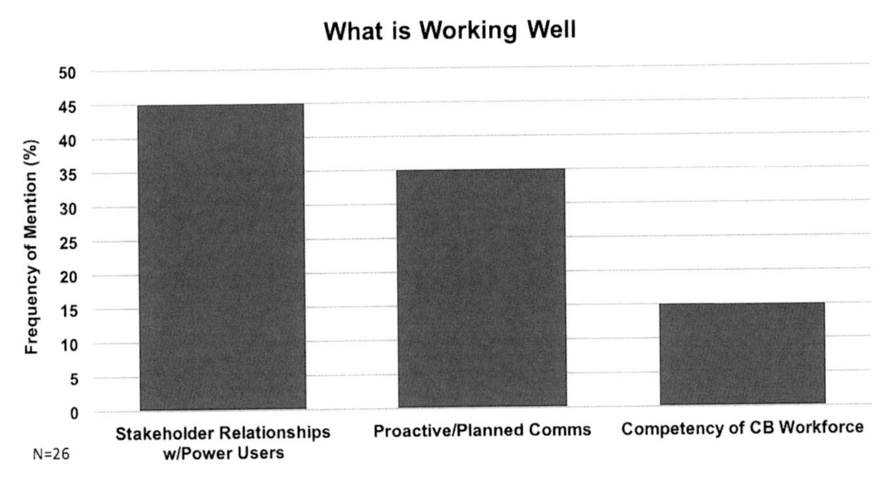


**Fig. 13.5** Currently effective forms of Census Bureau stakeholder engagement and communication

some stakeholders including Congressional staffers, and the expertise and ability of CB personnel to produce high quality data and data products that are timely and accurate (Fig. 13.5).

## Opportunities for Expanded Engagement

Interviewees indicated that there are opportunities for expanded engagement with certain stakeholders including: realtors, doctors, and other small business owners who could commercially benefit from increased sociodemographic knowledge about their communities; and the general public and teachers whose interests in such information may be more general, wanting to know more about the people in their neighborhood or community.

- "There is not a business in the country that would not benefit from the marketing data that we have. Most of them are paying for it right now; they are going to private firms. Every real estate agent should have an app on their iPhone and it should instantly know which school district I'm standing in and tell me the data that we have on the school district for the last 10 years. Every doctor who is looking to open up a practice wants to know the demographic makeup of the area that he's about to move into."
- "I'm constantly trying to paint the picture that making data for the public is a good idea and non-threatening and that these don't have to be academic products and that's okay. We can make a consumer product – that's who is paying us."
- "A great example is the education community, the K through 12. School teachers, every day, could probably benefit not just from incorporating the data into their classroom, but understanding where their kids are coming from or what the trends are going to be for next year for their classrooms. And you know we have all that data."

## *Current Challenges*

Interviewees identified a number of current challenges to stakeholder communications and engagement, including: inaccessibility of CB data including difficulty of finding and understanding CB data and data products, lack of stakeholder understanding of the value of CB data particularly of Congressional stakeholders who influence CB funding, CB's lack of understanding of stakeholders and their data needs, lack of stakeholder awareness of CB data, and increasing costs of CB data collection (Fig. 13.6).

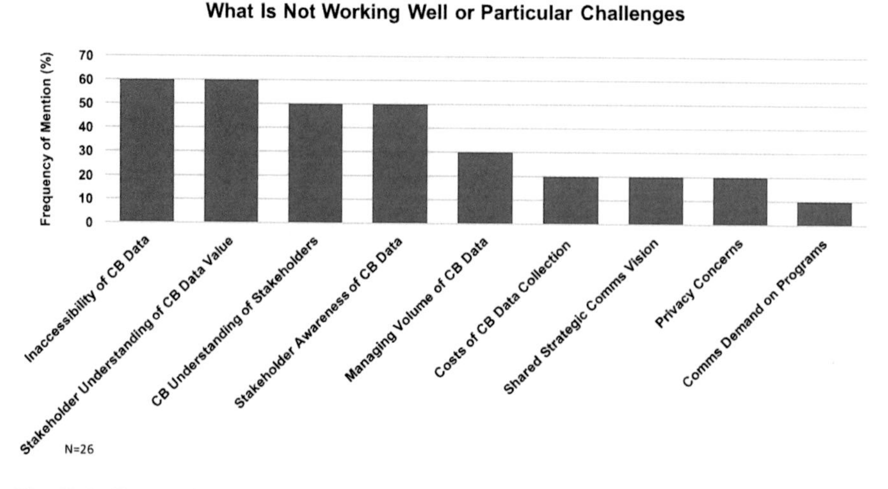


**Fig. 13.6** Current challenges for Census Bureau stakeholder communications and engagement

### Inaccessibility of CB Data

Most interviewees pointed out difficulties that stakeholders experienced in accessing and understanding CB data, in particular, difficulties in navigating the CB website and American FactFinder, an interface to Census data, to find the data that they were looking for and, once found, difficulties in using and understanding data due to its presentation format.

- "The primary mode of communicating the data that we have right now, which dwarfs everything in magnitude, is American FactFinder ... Any system that requires people to know what we know about the data before they can find the answer to their question is a bad system. There is nothing plain language about American FactFinder. That's the problem."
- "Ease of access. Our data is complex. Our data sets, our databases, are not easily understood. The American FactFinder system really, which is our major data tool for people to access huge amounts of data, discourages users. It is not simple to use. It's not easy."
- "The challenge is, how do I understand the data and, as I dig through your website, how do I not get lost, because there is so much there. I open the door and a million pieces of data coming flying at me like a Wizard of Oz bat thing."
- "I think the information could be somewhat intimidating for a small business to use. They just don't know where to start. A lot of businesses take our data and repackage it and sell it to businesses because we haven't done a good enough job of putting it in the right format that could be usable by any level of business."

### Lack of Stakeholder Understanding of the Value of CB Data

Most interviewees also raised concerns about the lack of stakeholders' understanding of the value of the data that CB collects and by association of the CB itself. This was highlighted as a particular concern with respect to Congressional stakeholders, who control CB budgets, and also survey respondents who make decisions about whether to respond to a survey or otherwise participate in data collection:

- "[The public] doesn't understand the importance of what the data, what those statistics we collect actually do for them in terms of helping not only ... elected officials ... make better decisions, but also businesses ... make a lot of decisions. So the information we provide helps drive economic development at the very lowest level, and without it, businesses would be floundering because they wouldn't know what is working. They wouldn't know who to market to. They wouldn't know what products to develop."

### Lack of CB Understanding of Its Stakeholders

Many mentioned that the CB doesn't understand its stakeholders and their specific data needs, particularly businesses and the public:

- "We need to find out how the outside world thinks about the stuff we are collecting, and we need to know how the public understands subjects that are of the public interest that cross over what we collect. Remember that what we collect is done for policy interests."
- "It is a hard balance for the Census Bureau to try to strike what the optimum set of products is, and in an attempt to please as many users as possible, there is almost a glut of products we produce. We should get a better handle on who is really using all these products, whether we need all to produce all these products, how are these products currently being used, is there a different set that might be better?"
- "The Census Bureau's basic modus operandi, of how we're organized into a demographic program and an economic program, is not the way that a lot of people think of using data. When you ask a person on the street what economic data is, they would include something like median household income. That's an economic measure to most people, and yet from our perspective, that's a completely different thing because the debt number is collected from people, not from businesses. And because of our own internal fiefdoms of how we've designed things, we tend to disseminate data the way it was collected, not necessarily the way it's used."

### Lack of Stakeholder Awareness of CB Data

Many interviewees mentioned that stakeholders, particularly casual data users such as the public, but also others such as businesses, are not aware of the CB data that is available and may be useful to them personally or for their business activities:

- "Publicizing surveys and making people aware of it is something we still have a long way to go to do well. I still meet statisticians who don't even know about the ACS even though they should be using it. We're nowhere near getting through to 100 % of the people who need to know about the data."

## *Improving Communication and Engagement*

When asked about improving stakeholder communications and engagement, interviewees offered the following suggestions: better aligning CB communications with stakeholder needs; adopting more strategic and planned

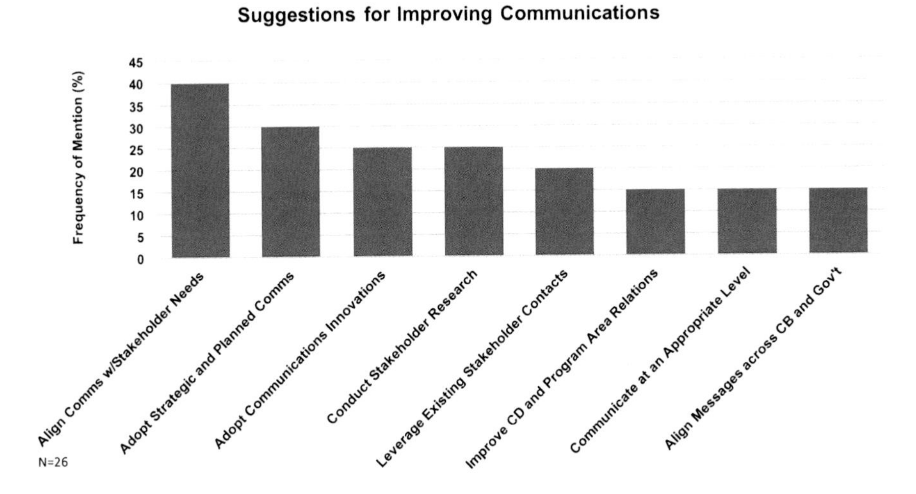

**Fig. 13.7** Interviewee suggestions for improving Census Bureau communications

communications processes, including developing a systematic, structured, strategic plan for communications that is proactive and aligned across the CB; adopting communications innovations, including new approaches to data collection, data products, and data visualization and taking advantage of new modes of communication and devices such as tablet computers, smartphones, and iPads; conducting stakeholder research to gain insight into how stakeholders interpret survey instruments, perceive the CB and its activities, and the ways that they could use CB data products to inform data collection, data products, and communications; and leveraging existing stakeholder contacts such as using contacts with ACS survey participants to test messaging for other data products such as the Decennial Census (Fig. 13.7).

## Preliminary Considerations on Key Components of the Communications Research and Analytics Roadmap (CRAR)

The mental models research results provided critical insight needed to develop the foundation for the Communications Research and Analytics Roadmap (CRAR) to support the Communications Directorate as it works to integrate communications services and improve participation in Census Bureau data collection and dissemination activities. Implementing the CRAR will enable the Census Bureau to design and implement an integrated communications program that is science based, customer centric, and data driven.

The research revealed the following value case, which we used as a fundamental driver for the design of the CRAR. A critical need, and a key driver for improving how the Census Bureau engages and communicates with funders, key stakeholders, and citizens was clearly articulated by one of the expert interviewees, who noted:

> "In 2000, the Census cost about $6.5 billion. In 2010 we are coming in at about $12.7 billion. If we make no changes in the Census design, the Census is going to cost close to $20 billion or more in 2020. Those escalating costs are untenable.... in 2010, we calculated that for every increase of one percentage point in response rates, we saved the government $82 million."

Others noted that the cost of Census Bureau data collection is growing while funding constraints are increasing. Census Bureau priorities, processes, methods, and tools are under increasing scrutiny. Interviewees discussed the need to:

- Increase data collection participation, while decreasing the cost;
- Increase the number of users of CB data, with emphasis on what they perceived to be "casual users" who could benefit significantly from the use of CB products and services;
- Improve user access to CB products and services through: enhanced internet access, specialized applications; better data visualization; more user-centric reports, etc.; and
- Enhance understanding of the value and application of CB data products to a wider range of users' daily life.
- Improve the collaboration and coordination of communications across CB to foster a more customer-centric focus.

Several interviewees believed that Census data collections could be made more cost effective by:

- Increasing response rates and decreasing need for follow-up; and
- Applying innovations such as the use of administrative records and expanding Internet response options.

## *Strategic Framework for CRAR*

The Foundational Research revealed a significant opportunity for the Census Bureau to broaden and deepen its user base by improving the use and perceived value of its products and services. The CRAR mission, goals, guiding principles, priority research questions, and research steps are illustrated in the Strategic Framework for the CRAR.[3]

---

[3] The detailed Mental Models Research Report: Key Internal Expert Stakeholders was provided to the Communications Directorate Project Team March 29, 2013.

## Strategic Framework for the Communications Research and Analytics Roadmap

**CRAR Mission**

**Develop the framework for CRAR and support its implementation in a way that:**

- Builds on the successes of the Census 2010 research program
- Addresses the primary research needs of the Communications Directorate in the nondecennial years
- Lays a strong foundation for the 2020 Census and delivers behavioral results

**CRAR Goal**

**Significantly increase the use and perceived value of Census Bureau, its products, and its services by all users and stakeholders through an integrated research approach focused on behavioral outcomes**

**CRAR Guiding Principles**

**Significantly increase the use and perceived value of Census Bureau, its products, and its services by all users and stakeholders through an integrated research approach focused on behavioral outcomes**

- Customer and user centric
- Collaborative
- Actionable and results driven
- Innovative
- Based on state-of-the-science methods and tools
- Built on existing Census knowledge and research
- Measureable, supporting iterative learning, and continuous improvement

**CRAR Priority Research Questions**

1. **Increasing Participation in Data Collection:** What should the Communications Directorate do to support increasing participation in data collection activities?
2. **Broadening the User Base:** What should the Communications Directorate do to engage and broaden its universe of data users?
3. **Improving the Value of Census Bureau Data Products:** How do users currently value Census Bureau data products and services? What should the Communications Directorate do to improve their use and enhance their value?
4. **Improving Internal Collaboration:** What should the Communications Directorate do to improve collaboration and foster a customer-centric focus throughout the Census Bureau?

**CRAR Research Steps**

**For each of the CRAR Priority Research Questions:**

**Step 1:** **Integrate Expert Insight and Existing Research:** Integrate expert insight and existing research to develop hypotheses and potential approaches to address the Priority Research Questions in a systematic, structured way

**Step 2:** **Conduct Stakeholder and Audience Behavioral Research:** Test hypotheses through specific behavioral qualitative and/or quantitative methodologies that identify influences on decision making and action

**Step 3:** **Develop and Refine Strategies, Messages, Content, and Data Products:** Develop and refine communications strategies, messages, and data products to influence stakeholder and audience decision making and behavior, using an iterative approach so studies build upon one another

**Step 4:** **Test and Evaluate:** Conduct field tests of strategies, messages, content, and data products to ensure measurable, real-world performance. Implement key performance measures and analytics to identify opportunities for further improvement

**Acknowledgement** Based on the Executive Summary: "Census Bureau Integrated Communications Services for Data Dissemination: Mental Models Research with Key Internal Expert Stakeholders" (May 2013), prepared for the Census Bureau Communications Directorate by Decision Partners with support from Reingold and Penn Schoen Berland. Special thanks to contributors: Monica Wroblewski of Census Bureau, Joseph Ney of Reingold and Robert Green and Samuel Hagedorn of Penn Schoen Berland, for their assistance with this chapter.

# Part IV
# Mental Modeling Software Support

# Chapter 14
# Supporting and Expanding the Scope and Application of Mental Modeling: Current and Future Software Developments

**Daniel Kovacs, Alex Tkachuk, Gordon Butte, and Sarah Thorne**

## Introduction

This ongoing development of software platforms enables application of the Mental Modeling approach by a broader group of practitioners applied to a wider array of risk and decision topics. The IDST™ and RiskLogik™ platforms take advantage of new information technologies and ensure that Mental Modeling remains a state-of-the-science approach for addressing complex challenges and engaging clients, experts, stakeholders, and researchers using computer-based technologies.

## CASS Support Software for Mental Modeling Technology™ (MMT™) Research Processes

Decision Partners' developed CASS, the integrated Cognitive Analysis Software Suite that is specifically designed to efficiently support the unique empirical methods embodied in the research component of MMT™ from developing graphical

D. Kovacs, Ph.D.
Decision Partners, 1458 Jersey Street, Lake Milton, OH 44429, USA
e-mail: dkovacs@decisionparters.com

A. Tkachuk (✉)
Decision Partners, 88 Marksa Street, Apt. 258, Obninsk, Kaluga Reg. 249030, Russia
e-mail: atkachuk@decisionpartners.com

G. Butte
Decision Partners LLC, Suite 200, 313 East Carson Street, Pittsburgh, PA 15217, USA
e-mail: gbutte@decisionpartners.com

S. Thorne, M.A.
Decision Partners, 1084 Queen Street West, #32B, Mississauga, ON, Canada, L5H 4K4
e-mail: sthorne@decisionpartners.com


M.D. Wood et al., *Mental Modeling Approach*, Risk, Systems and Decisions,
DOI 10.1007/978-1-4939-6616-5_14

depictions of the systems being modeled, through coding, analysis of qualitative data, back to graphical depiction of research results. CASS was developed to enable researchers to hypothesize, visualize, tabulate, analyze, and report on individuals' mental models of complex social and technical issues. It provides the analytical framework for conducting and analyzing mental models research.

CASS has enabled significant research process and productivity improvements and provides the platform for Decision Partners' Mental Modeling Technology™. To date Decision Partners has developed two components of CASS: eCASS and cCASS. eCASS enables the development of visualizations, primarily expert models, mental models, and stakeholder maps. As mentioned in Chap. 13, the expert model serves as the analytical structure for analysis and reporting of mental models research. Stakeholder maps enable the visualization, characterization, and prioritization of the universe of stakeholders for a specific topic or issue. The complementary cCASS software supports the coding and analysis of mental models research data. When the mental models research results are visualized by creation of color-weighted models, CASS provides end-to-end support for the MMT™ research process.

The following provides an overview of the key features of eCASS and cCASS.

## *eCASS Software for Modeling*

Modeling is an essential part of the MMT™ approach. An expert model summarizes the relevant knowledge about the complex issues comprising the system to be managed and helps to identify and characterize issues of importance to stakeholders and factors that influence their perceptions, knowledge, and decision making. Expert models identify the core issues for action including options for strategies and communications needs and serve as the analytical framework for mental models research. When done well, expert models provide a strong, flexible framework for obtaining systematic assistance from experts, as well as documenting the assumptions underlying information. They allow for an effective and respectful way for communicators and technical experts to ensure that they have a common frame of understanding.

The MMT™ model drawing module, eCASS, shown in Fig. 14.1 enables visualization of several different types of models, including: expert models, mental models, and other types of influence or relational diagrams such as concept maps, which illustrate the relationships between concepts, and stakeholder maps.

A key feature of eCASS is the ability to organize multiple models into a structured project library (see Fig. 14.2) enabling the organization and multiple models that show various perspectives within each project (e.g., experts, lay stakeholders, or specific groups of stakeholders) or focus on specific aspects of the models at higher levels of detail (e.g., submodels that focus on stakeholder communications and engagement, or desired outcomes). eCASS also has some interactive features, such as the ability to "link" nodes and models, enabling the user to move from one model to another by clicking on specific nodes within a model.

Another key feature of eCASS is the ability for the user to organize variables, the elements of the model depicted as nodes, in a hierarchically structured variable

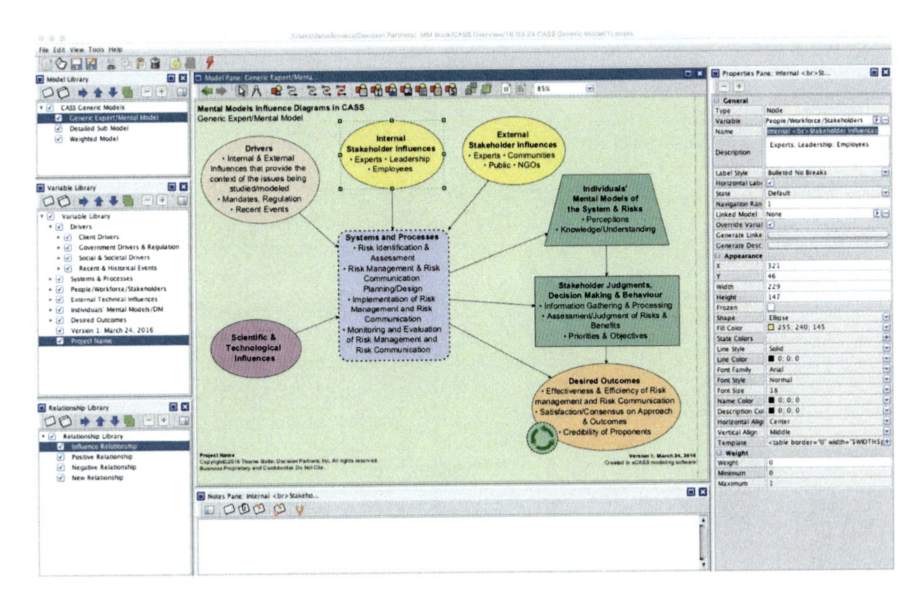


Fig. 14.1 eCASS application space

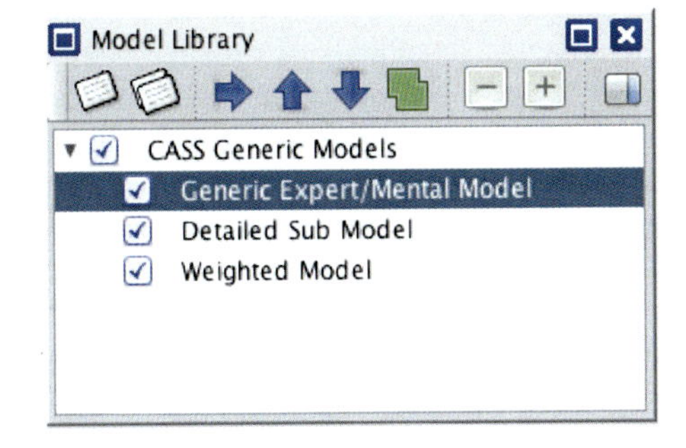


Fig. 14.2 eCASS model library

library (see Fig. 14.3). This enables the user to logically group similar variables and to have similar formatting of related variables both within and across projects. Variables from the library can be dragged onto specific models within a project to create a node or "instance" of the variable in that model, usually an oval, that allows depiction of the influences to and from that variable within the system. The node will have the default style associated with that variable (e.g., size, shape, color), but these features can be customized for each model, overriding the default formatting to allow for highlighting or weighting of nodes (e.g., to illustrate the results of research, see Fig. 14.4). Having default formatting for variables speeds model development and aids visual consistency among models, while allowing for customized formatting enables depiction of a broad range of models.

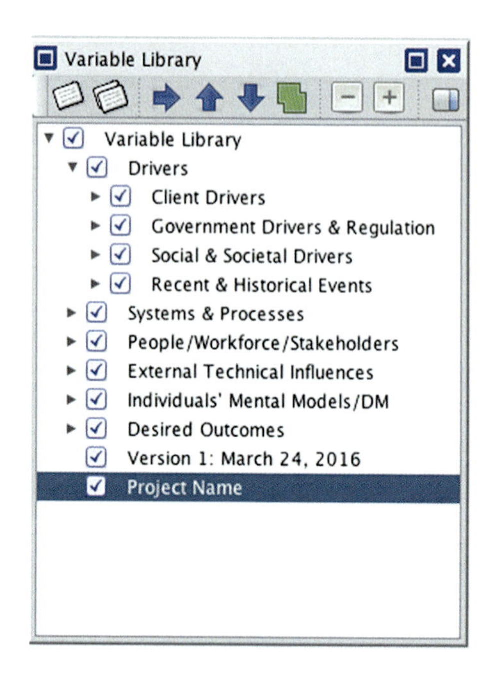


**Fig. 14.3** eCASS variable library

Relationships among variables are illustrated by connectors that link nodes that persist even when nodes are moved within a model (enabling more efficient editing of models). Relationships between nodes are created by clicking on the center of one node and dragging the cursor to the node to be linked. Different types of relationships can be illustrated by different types of connectors (e.g., dashed lines or two-way arrows), which can be created by customizing the formatting of the individual connector or by defining multiple types of connectors that can be selected from a relationship library (see Fig. 14.5).

Several preformatted model templates are included in eCASS to increase the efficiency of model development and consistency across models. These templates include a default variable library structure that includes the elements (variables) most commonly depicted in expert and mental models (see Fig. 14.6), such as:

- Project or system drivers.
- Systems or process elements.
- Human influences including employees and stakeholders.
- Communication and engagement processes and influences.
- Science and technology influences.
- Mental models, including perceptions, judgments, and decision making.
- Desired outcomes.

The predefined templates and variables can be customized for specific types of projects. The models and variables in the template can then be edited to develop models relevant to the specific project.

Other useful features offered in eCASS, include the ability to link to or attach external materials (documents, images, web links, etc.) to model nodes to create a

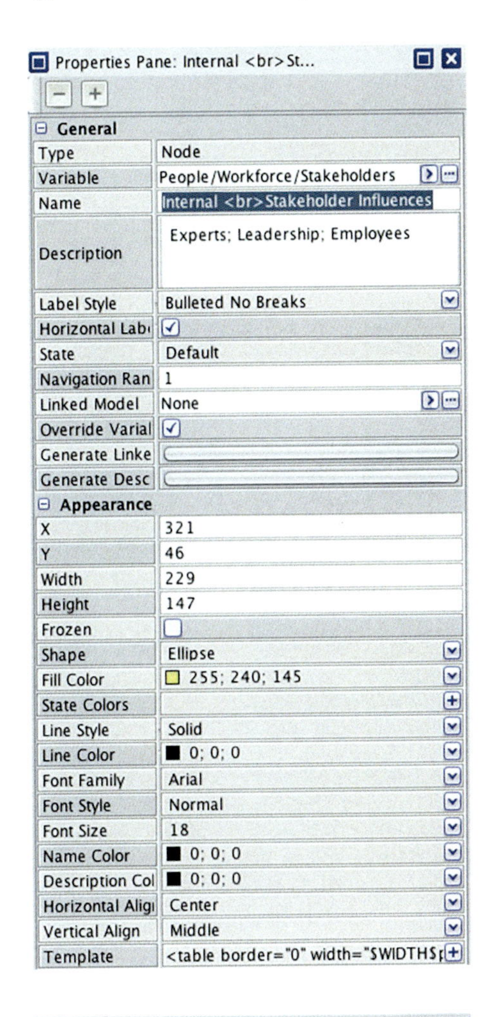


**Fig. 14.4** eCASS properties pane

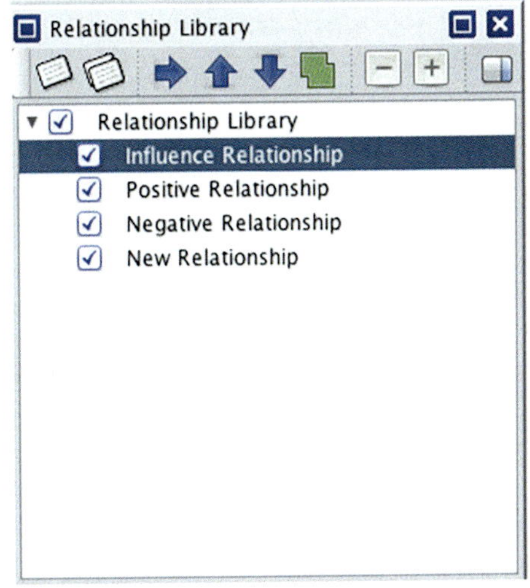


**Fig. 14.5** eCASS relationship library

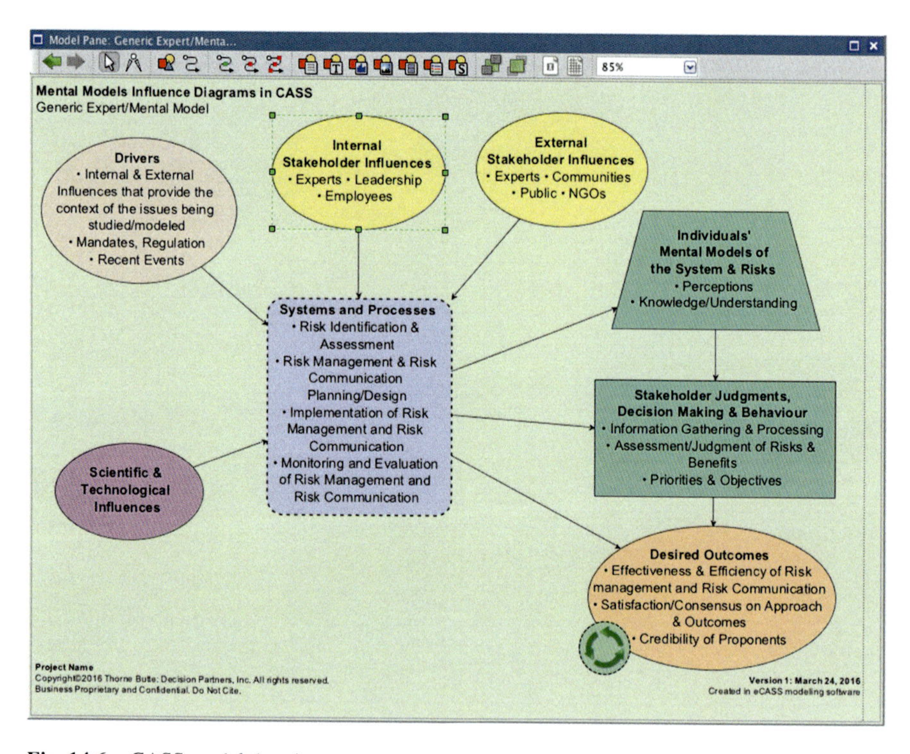


**Fig. 14.6** eCASS model drawing pane

knowledge repository or interface. In addition, all model elements (models, variables, nodes) can be annotated to enhance alignment and collaboration among those working on or reviewing models, and supporting reporting and narration of models.

## *cCASS Coding and Analysis Module*

Individuals' mental models are the tacit webs of belief that guide learning, interpretation of information and the complex psychological processes of judgment, decision making, and behavior on complex topics with social and technical influences. As such, they can only be uncovered through the semistructured qualitative research that is the core of the mental models research approach. The approach probes, in-depth, not just *what* people think, but *why* they think what they do. Exploring, understanding, and characterizing these mental models requires a structured coding and analysis process that is grounded in the analytical framework of the expert model while allowing for unanticipated concepts and themes to emerge. Specific methods and tools are required to explore, understand, and characterize insight from the qualitative data collected by the exploratory mental models research approach.

The MMT coding and analysis module, cCASS, enables coding and analysis of qualitative and quantitative data collected in mental models research (see Fig. 14.7

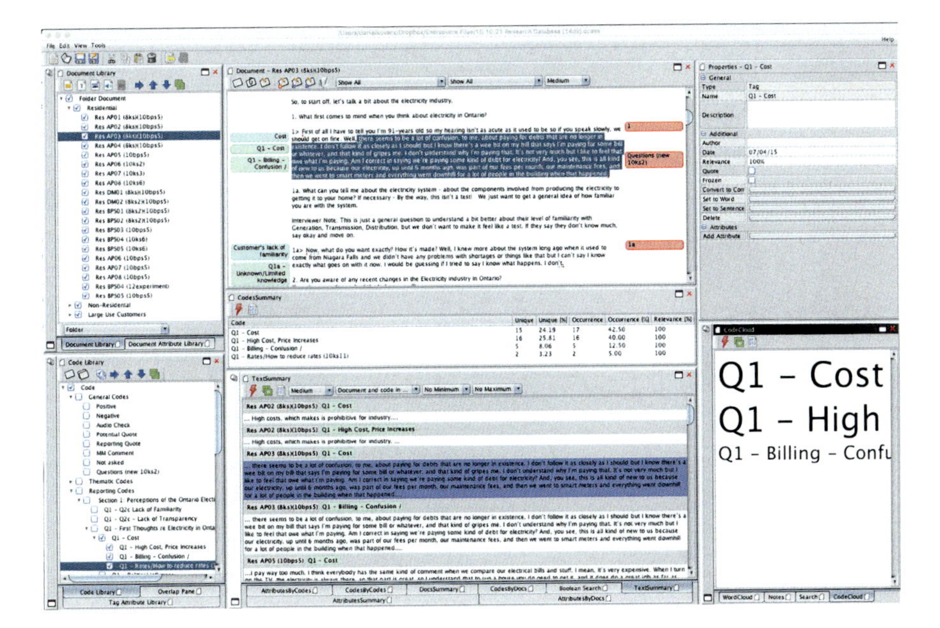


**Fig. 14.7** cCASS Application Space

cCASS Application Space). This data is primarily text-based transcripts of one-on-one in-depth interviews. When formatted appropriately some automatic coding is applied to the imported files such as application of question number codes and coding of structured ratings questions. This speeds the coding and tabulation of such data and enables exploration of the qualitative data by allowing for comparison of responses across interviews. This significantly improves the ability of the analyst to identify emerging themes, which is a core benefit of the MMT approach, as it is often applied to problems that are not well understood or fully characterized.

Interview transcripts are then imported and grouped by cohort (e.g., primary stakeholder groups) to enable the analyst to assess the gaps and alignments on key topics between the research cohorts. In addition, demographic data is also automatically coded and applied at the document level to allow for comparisons of subgroups (e.g., by gender or stakeholder age; see Fig. 14.8 Document Library).

Relevant variables created during development of the expert model can be imported into the cCASS code library. Additional codes can be created as needed during coding to represent new topics or more complex themes that characterize individual's mental models. Variables can be organized into a logical structure that promotes reproducible and reviewable coding processes (see Fig. 14.9).

Codes are then applied by highlighting a segment of a document (see Fig. 14.10). This is done by a research analyst who has been briefed on the project and trained to in analysis techniques to enable identification of themes (known and emerging). Once applied, these codes can be used to further explore the data in a "parallel coding" process where the data is "filtered" to show similar responses across all interviews that have the same question number or topical codes. This allows for further

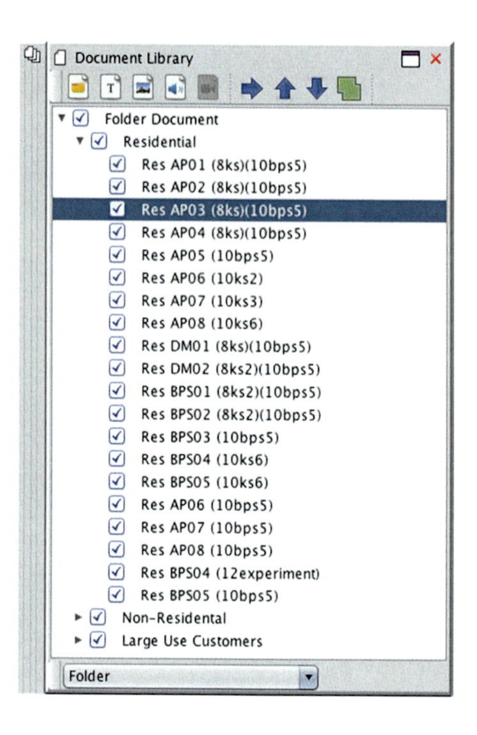


Fig. 14.8 cCASS document library

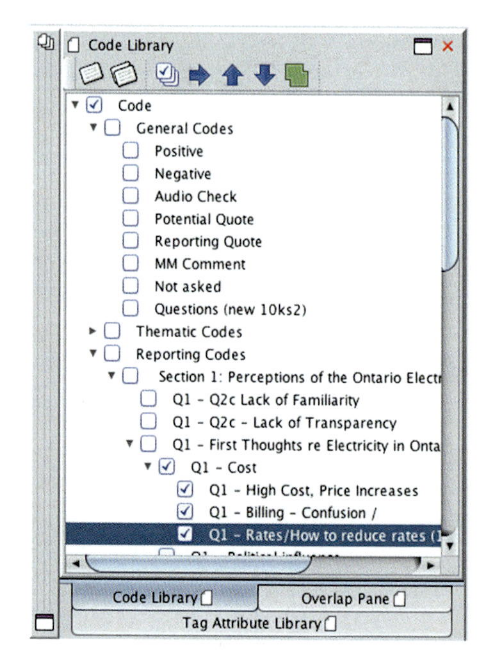


Fig. 14.9 cCASS code library

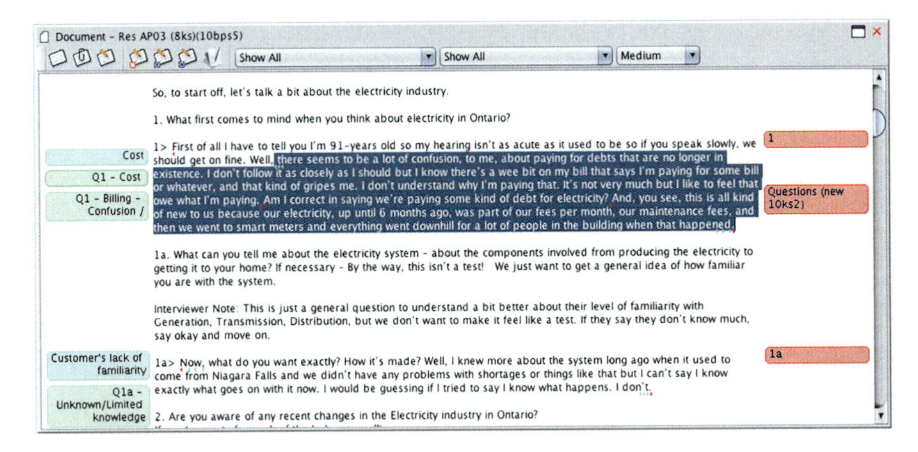


**Fig. 14.10** cCASS document pane with codes applied

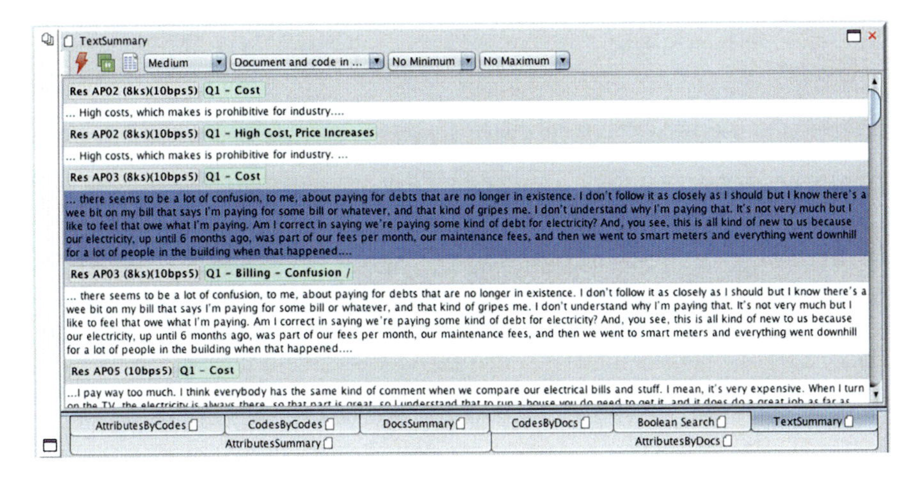


**Fig. 14.11** cCASS text summary pane showing filtered responses across interviews

refinement of the analysis (see Fig. 14.11). This iterative, "parallel coding" process can be more effective and efficient than one-pass (or multiple-pass) "linear coding" where an interview is coded from beginning to end before going on to the next interview. This is again particularly appropriate for mental models research data where exploration of the data is needed to gain the in-depth insight needed to characterize individuals' mental models.

The cCASS software platform is designed to allow for the work of multiple coders operating in parallel. Multiple coders can be assigned to code the same or different segments of the research interviews. The databases of the respective coders can then be merged to produce a combined coded document for subsequent tabulation and analysis. The ability to merge databases also allows for review and feedback of coding (e.g., by fellow coders or by a lead coding manager) by enabling changes and com-

CodesSummary

| Code | Unique | Unique [%] | Occurrence | Occurrence [%] | Relevance [%] |
|---|---|---|---|---|---|
| Q1 - Cost | 15 | 24.19 | 17 | 42.50 | 100 |
| Q1 - High Cost, Price Increases | 16 | 25.81 | 16 | 40.00 | 100 |
| Q1 - Billing - Confusion / | 5 | 8.06 | 5 | 12.50 | 100 |
| Q1 - Rates/How to reduce rates (10ks11) | 2 | 3.23 | 2 | 5.00 | 100 |

**Fig. 14.12** cCASS code summary pane showing tabulation of code frequency

ments in coded documents to be sent back to the coder for incorporation into their future coding.

The research analysts can then use the cCASS software functionally to tabulate the frequency of mention of each of the relevant codes representing specific themes across all interviews or within specific cohorts or subgroups (see Fig. 14.12).

The platform also supports reporting of MMT results by generating tabulated numerical results and exemplar quotes associated with qualitative mental models analysis. Quantitative data can be copied into MS Excel for generation of charts and text-based illustrative quotes can be copied to MS PowerPoint or MS Word for reporting.

## *CASS Module Integration (eCASS and cCASS)*

Modules in the CASS are integrated to enable sharing of elements between eCASS and cCASS (and with other word processing and presentation software) to further enhance the efficiency and consistency of the research process across research steps. For eCASS, the variable library can be exported as text to facilitate narration of models in reporting and as the analytical framework to facilitate development of the research interview protocol. The variable library can also be exported directly into the cCASS coding library to facilitate and align coding and analysis with the analytical Expert Model Framework. The models produced in eCASS can be exported as graphical images in a number of formats to facilitate reporting and presentation of research results. Such integration can improve the overall efficiency of the research process by reducing the need to reenter data and descriptions in multiple places.

## *CASS Development*

The CASS platform is being continuously developed to improve functionality and usability. Recent exploratory functionality added to cCASS includes the ability to tabulate survey-type quantitative data from a large number of respondents along with associated unstructured, open-ended responses. This hybrid research approach blends the representative power of quantitative surveys with the exploratory power of qualitative mental models research. The cCASS software also has some functionality to enable coding of other materials such as unstructured text documents (e.g., news articles and other web-based materials), as well as audio and video files that represent new sources and modes of collection of research data that are becoming more available. Additional exploratory functionality is being investigated that will further integrate cCASS and eCASS. Analysts will be able to transfer quantitative results from cCASS back to eCASS to enable automatic generation of weighted models to illustrate research findings.

CASS is a platform that is continuously been developed and updated to meet the needs of MMT and improve usability. For more information about CASS, contact Decision Partners at: dprc@decisionpartners.com.

## Case Study: IDST™ Used by Enersource Hydro Mississauga to Fulfill Customer Engagement Regulatory Requirements

### *The Customer Engagement Challenge*

In 2015 Enersource Hydro Mississauga became the first organization to deploy a new high-tech, mental models-based approach using Interactive Decision Support Technology™ (IDST™) for engaging customers. The following summarizes the need for and implementation of this new technology, along with the results.

In early 2013, Ontario's energy regulator, the Ontario Energy Board, implemented new requirements for rate application submissions calling for electricity distribution companies to solicit their customers' input and demonstrate that their Distribution System Plan (DSP) is responsive to customers' input.[1] The DSP is the foundation for rate increases over the next 5 years. While rate increases are a topic of great interest to most customers, the plan behind them and its rationale have not been transparent to most in the past.

Given the complex social and technical issues inherent in the DSP, the Enersource Team determined that effective customer consultation had to be based on in-depth insight into people's values, interests, and priorities, in short, their mental models.[2] Decision Partners designed and conducted mental models interviews with three cohorts of customers: residential, nonresidential, and large use customers.

This foundational research demonstrated that most residential and nonresidential customers did not have a sufficient understanding of the electricity system to provide meaningful input into the DSP Distribution system plan (DSP). Customers' mental models of the electricity system were incomplete and, in some cases, incor-

---

[1] http://www.ontarioenergyboard.ca/oeb/_Documents/Regulatory/Filing_Requirements_Tx_Dx_Applications_Ch5.pdf.

[2] For example: Johnson-Laird, P.N. (1983). Mental Models. Cambridge, MA: Harvard University Press. Atman, C.J., et al. (1994) Designing risk communications. Risk Analysis 14(5): 779–788; Bostrom, A. et al. (1992) Characterizing mental models of hazardous processes. J. Social Issues 48(4): 85–100; Fischhoff, B. et al. (1997) Risk perception and communication. In: Detels, R. et al. (eds.) Oxford textbook of public health. London: Oxford University Press, 1997. Pp. 987–1002.

rect. Consequently, if customers did not understand the system, they could not reasonably be expected to understand the rationale of the investment decisions being proposed in the DSP or provide meaningful input on the key components.

The foundational research also underscored that customers were unclear about the company's role in the electricity system and that there was significant confusion about the costs associated with Enersource's distribution compared to the costs of the overall electricity system. Consequently, Enersource's value proposition—what it takes to deliver electricity safely and reliably—was not understood by most customers. An intervention was required.

## *New Technology for Customer Engagement*

Decision Partners partnered with MedRespond[3] to develop a new web-based system for stakeholder engagement using "Interactive Decision Support Technology™ (IDST)." Combining cognitive modeling and artificial intelligence to engage customers and elicit input, IDST™ integrates Decision Partner's Mental Modeling Technology™ with MedRespond's Synthetic Interview® artificial intelligence and online communication products.

The Synthetic Interview® is an interactive technology that simulates a one-on-one conversation through a combination of streaming video and artificial intelligence designed to simulate a direct interaction with the video hosts. Customers were able to listen to and watch individuals describe various critical topics and are then asked for their feedback—all in real time. The content, built on the foundational mental models research results, was designed to address specific gaps identified in customers' understanding of the topic being studies, while enabling participants to provide meaningful input on the various topics. They could also ask questions to learn more about specific topics that interest them (Fig. 14.13).

## *IDST™ Experience*

Enersource's IDST™ site—www.electricitydialogue.ca[4]—was launched November 23, 2015 and remained active until January 7, 2016 (Fig. 14.14).

The IDST™ enabled customers to work through the various DSP components and the rationale for investment in it. The presence of an Enersource host, commenting or asking questions along the way, gave them the opportunity to think through the benefits, risks, and trade-offs of the proposed activities and the level of investment being recommended. This in turn enabled customers to make *well-informed judgments* of the acceptability of the recommended investments.

[3] MedRespond (www.medrespond.com) is an online communications company that combines artificial intelligence, search and streaming media to enable enterprises to provide interactive and personalized communication for their customers, patients, and clients.

[4] This the site is no longer active, but a summary of the IDST can be seen on the YouTube: https://youtu.be/tc4iMeZHx8Y.

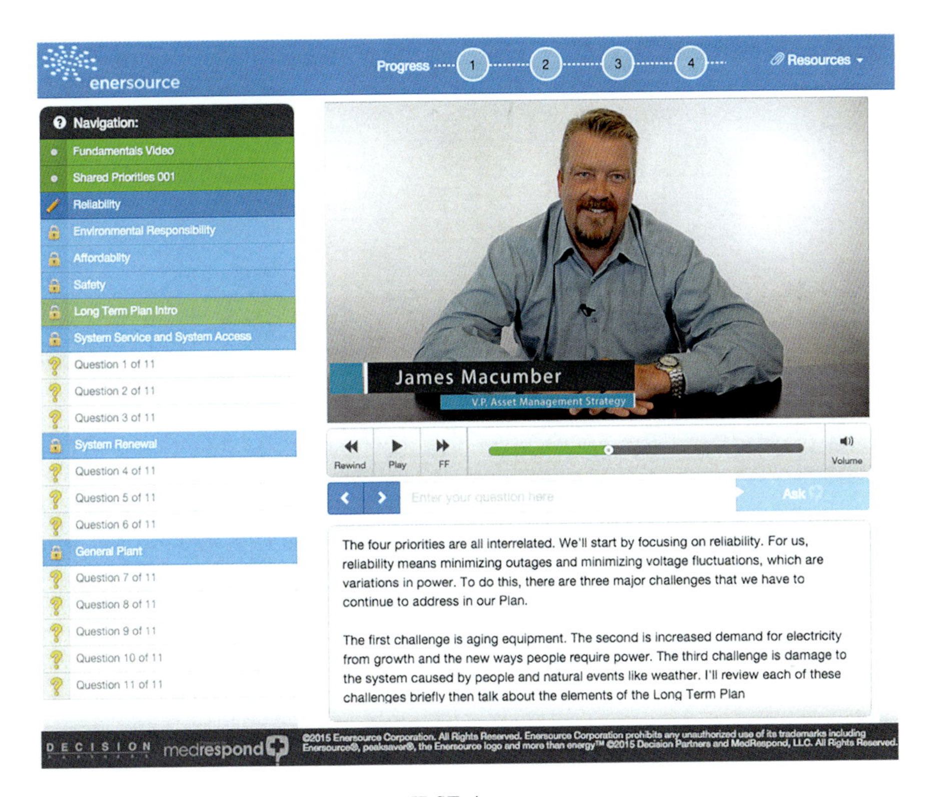


**Fig. 14.13** Screen capture of Enersource IDST site

## *The Results*

For Enersource, the Customer IDST™ program delivered far more value than traditional approaches and the quality and value of the customer engagement was significantly higher and the outcome more robust:

- 2157 Residential Customers and 49 Non-Residential Customers visited the site—more than double the response achieved by an electricity distribution company of a similar size to their online workbook landing page.
- 1358 Customers completed the DSP survey of the IDST™—more than a tenfold increase over the workbook completion of the three electricity distribution companies whose workbooks we assessed.
- The average engagement time was 25.7 min per visitor—far less than the time required to complete a workbook or attend a focus group meeting, but sufficient time to think through the components of the DSP and provide meaningful input.

The following chart summarizes customer engagement on the site (Fig. 14.15).

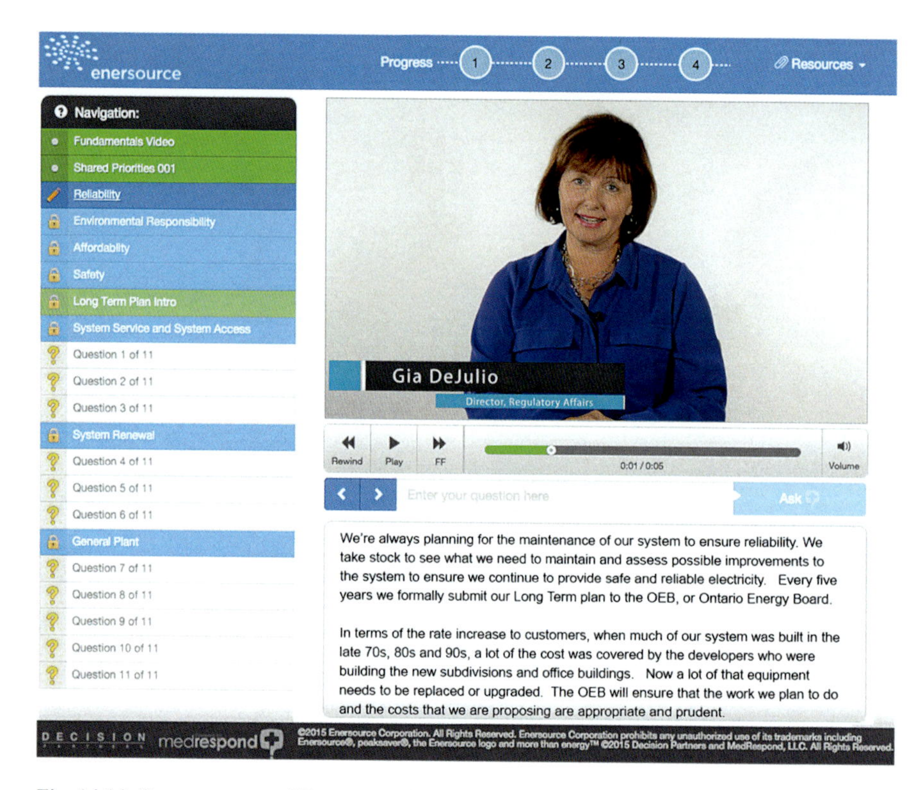


Fig. 14.14 Screen capture of Enersource IDST site

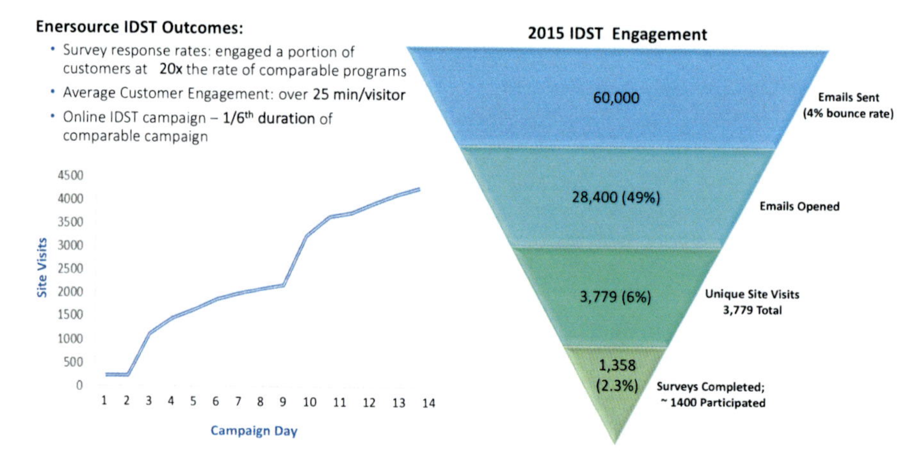


Fig. 14.15 Enersource IDST campaign summary

### *Advantages Over Conventional Customer Engagement Methods*

IDST is a more customer-centric approach than traditional engagement tools because it is based on individuals' mental models—*their* values, interests, priorities, and information gaps and needs. The technology then anticipates and responds to these in a conversational format. This dialogue-based technology enhances stakeholder engagement through the proven Synthetic Interview® medium that enables stakeholders to gather the information they want and need in order to make well-informed judgments. At the same time, it enables real time, systematic collection of data about stakeholder interests and priorities and can assess how these change over time—in short, how their mental models evolve over time as new and relevant information become available to them.

For more information, see the YouTube of the Enersource Case Study https://www.youtube.com/watch?v=tc4iMeZHx8Y&feature=youtu.be, or contact us through www.decisionpartners.com or www.medrespond.com.

### *Considerations on Future Applications of the IDST™*

The first full-scale application of the IDST™ for customer engagement demonstrated its potential to reach out to and engage with a broad spectrum of stakeholders on a complex topic. Near-term opportunities could include broad stakeholder engagement and ongoing mental model-based dialogue and data collection on challenges ranging from siting major energy projects such as pipelines, liquefied natural gas (LNG) facilities, and transmission lines, to engaging patients, parents, caregivers, and healthcare providers in dialogue on complex topics such as vaccine safety, public health implications of Ebola and Zika virus, and beyond.

The IDST™ enables project and program managers, risk managers, and communicators to first address their stakeholders' mental models of the topic or issue at hand—what they know, don't know, misunderstand, and want to know—through a user-friendly interface. Over time, they can track and analyze the changing mental models of IDST users. New interests and priorities, along with gaps in understanding, can be identified enabling the IDST sponsor to refine existing modules or create new ones. The IDST can also be used to conduct training, elicit expert or layperson input, test scenarios, and so on.

## Mental Modeling Technology™ with Quantitative Risk Analysis Tools

In cases where the client requires a quantitative picture of risk, networked risk analysis tools, such as RiskLogik,[5] can be integrated as an extension of the CASS tool suite. RiskLogik is a direct analysis tool for understanding, mapping, and mitigating

[5] See: www.risklogik.com.

complex risk such as critical infrastructure dependencies in cities or component dependencies in nuclear power plants and chemical facilities. Using graph theory techniques, RiskLogik creates directed graphs that discover the pathways of exposure to risk, the potential systems effect and any cascades of consequence that might emerge from disruptive events occurring in those cities or at such sites. The RiskLogik toolset can calculate the cost of disruptive events, the cost of mitigations, and both the geospatial influences and consequences of either. For complex sites RiskLogik can do this in 3D using constructive simulation tools.

While the base data for RiskLogik models is usually collected using Delphi technique, MMT™ supported by eCASS represents an expert data elicitation and visualization capability for complex models that is both rigorous and auditable aligning it perfectly with the practice requirements for resilience engineering. Conversely, the computive power of RiskLogik models provides quantitative analysis to eCASS models rendering them dynamic and auditable in multiple domains.

The connection to eCASS occurs when stakeholder models and/or expert models are used to inform the values in a RiskLogik quantitative risk model. Using standard statistical analysis methodology the weighting of nodes and relationships in the eCASS models are imported to the RiskLogik model and used as base data to drive the strongest path algorithms. The result in either case is a dynamic quantitative model of the risk environment that is geo-enabled, is quantifiable, is auditable and repeatable, that can be costed, that can be evolved over time to reflect changing circumstances and understanding, and that can be interrogated to determine optimum risk mitigation strategies.

For the complex risk analysis required to understand the potential impacts of climate change on industries, economies, and civil societies, the combined technologies, called Resilience Solutions Technologies™ (RST™) has, as yet, no peer.

RST™ is a state-of-the-science integration of three elements: Mental Modeling Technology™ for understanding people's judgment, decision making and behavior; advanced risk analysis, cyber resilience, geospatial analysis, and constructive simulation tools in the RiskLogik™ toolset; and, world leading resilience analysis methodology based on academic research. RST™ is delivered through a seven-stage service delivery model that engages all the appropriate elements of the client's organization in the discovery, analysis, and mitigation of resilience risk, followed by ongoing support to implement the required resilience measures and ensure ongoing adaptation as the hazard environment changes over time. Organizations can elect to adopt as much or as little of the methodology and technology as they wish at any stage in the engagement process.

**Acknowledgments** Special thanks to Katherine Sousa and Alexander Tkatchuk of Decision partners, Gia DeJulio of Enersource Hydro Mississauga, Virginia Pribanic and Carlton Ketchum of MedRespond, and Nick Martyn of RiskLogik for their contributions to this chapter.

## References

www.decisionpartners.com
www.medrespond.com
www.risklogik.com

# Index





M.D. Wood et al., *Mental Modeling Approach*, Risk, Systems and Decisions,
DOI 10.1007/978-1-4939-6616-5

**W**